Mobile Technologies for Every Library

Medical Library Association Books

The Medical Library Association (MLA) features books that showcase the expertise of health sciences librarians for other librarians and professionals. MLA Books are excellent resources for librarians in hospitals, medical research practice, and other settings. These volumes will provide health care professionals and patients with accurate information that can improve outcomes and save lives. Each book in the series has been overseen editorially since conception by the Medical Library Association Books Panel, composed of MLA members with expertise spanning the breadth of health sciences librarianship.

Medical Library Association Books Panel

About the Medical Library Association

Founded in 1898, MLA is a 501(c)(3) nonprofit, educational organization of 4,000 individual and institutional members in the health sciences information field that provides lifelong educational opportunities, supports a knowledgebase of health information research, and works with a global network of partners to promote the importance of quality information for improved health to the health care community and the public.

Books in Series

The Medical Library Association Guide to Providing Consumer and Patient Health Information, edited by Michele Spatz
Health Sciences Librarianship, edited by M. Sandra Wood
Curriculum-Based Library Instruction: From Cultivating Faculty Relationships to Assessment, edited by Amy Blevins and Megan Inman
Mobile Technologies for Every Library, by Ann Whitney Gleason

Mobile Technologies for Every Library

Ann Whitney Gleason

ROWMAN & LITTLEFIELD
Lanham • Boulder • New York • London

Published by Rowman & Littlefield
A wholly owned subsidiary of The Rowman & Littlefield Publishing Group, Inc.
4501 Forbes Boulevard, Suite 200, Lanham, Maryland 20706
www.rowman.com

Unit A, Whitacre Mews, 26-34 Stannary Street, London SE11 4AB

Published in cooperation with the Medical Library Association

British Library Cataloguing in Publication Information Available

Library of Congress Cataloging-in-Publication Data

Gleason, Ann Whitney, 1960–
Mobile technologies for every library / Ann Whitney Gleason.
pages cm. — (Medical Library Association books)
Includes bibliographical references and index.
ISBN 978-1-4422-4891-5 (hardback : alk. paper) — ISBN 978-1-4422-4892-2 (pbk. : alk. paper) 1. Mobile communication systems—Library applications. I. Title.
Z680.5.G58 2015
025.0422—dc23
2014046357

∞ ™ The paper used in this publication meets the minimum requirements of American National Standard for Information Sciences Permanence of Paper for Printed Library Materials, ANSI/NISO Z39.48-1992.

Printed in the United States of America

Contents

Preface

Mobile technology is everywhere. You see it on buses, at the beach, while walking on city sidewalks, and ubiquitously in most public spaces. The Horizon Report for 2013 rated tablet computers as one of the two top emerging technologies in higher education. Gartner Research listed mobile apps as one of the top strategic technology trends for 2014. In the summer of 2014, it was reported that there were 1.2 million active apps in the Apple Store. At the same time, Android reported over 1.3 million available apps. Hobbies, health, beauty, science, nature, travel; there is an app for everything now. Businesses use apps to reach out to new markets and customers. Thousands of apps are added monthly to the Apple Store and Google Apps Marketplace. The time to watch and wait for mobile technology to mature is over; now is the time to adopt mobile technology to enhance your library programs and services.

If you are wondering what mobile technology adoption means for you or how to get started, this book will answer your questions in detail. Wondering about the opportunities and pitfalls of mobile technology use in libraries? This book will answer your questions. Thinking of starting a mobile program in your library? Want to improve on existing services or add new ones? This book will address questions about platforms, options, security, best practices, and more. You will find previews of many useful apps for libraries. Web links and resources for further research are also included.

Mobile Technologies for Every Library is organized by chapters on specific topics related to mobile technology, including the history of mobile technology, existing types of mobile technologies, popular mobile devices, and supporting technologies. Ways to provide mobile technology for your users, a survey of currently available apps of use to libraries, ways to use mobile technology for library work, best practices, and future directions are discussed in detail. Each chapter is organized by subtopics with tips and examples from real library programs to help you get started with using mobile technology in your library.

Chapter 1 gives background information about mobile technology and provides a concise history of mobile devices. Chapter 2 gives detailed information on the popular mobile devices in use today, as well as information about supporting technologies such as wireless networking. Chapter 3 presents information on making the library website more accessible to mobile users. Chapter 4 discusses apps for libraries and covers app development for those who are interested in creating a library app. Chapter 5 presents the ways that people use mobile technology to access information, the types of information searched for, and what that means for libraries. Chapter 6 lists by category selected apps that are particularly useful to libraries. Chapter 7 gives guidelines and best practices for a variety of mobile technology programs in libraries. Chapter 8 presents ideas for using mobile technology in library education. Chapter 9 provides ideas for doing library outreach and communication with mobile technology, and chapter 10 discusses future trends and directions.

Read on and prepare to welcome mobile technology into your library!

Chapter One

Background and History of Mobile Technology

Networking technology company Cisco Systems, Inc. stated in a September 2013 article that mobile Internet is the top disruptive technology likely to change our lives in the next decade.[1] A disruptive technology is defined as one that creates a new market for consumers and goes on to displace former technologies. The Cisco article proposed that with the rapid growth of the mobile Internet, the majority of people not previously connected to the Internet will connect through mobile devices. Also in September 2013, Pew Research Center reported in its "Cell Internet Use 2013" report that almost two-thirds of cell phone owners surveyed go online using mobile phones.[2] There are over 300 million cell phone users in the United States alone. Clearly, mobile technology is a game-changing innovation. The implications of this rise in mobile access to the Internet are huge for businesses wanting to reach future mobile markets. Libraries also need to take the mobile revolution seriously and plan for services to mobile users if they wish to remain relevant in an increasingly mobile world.

Most libraries have existing websites giving patrons access to library information, resources, and services online. Some libraries have created mobile-friendly websites but few provide full mobile access. The American Library Association's "State of America's Libraries Report 2013" notes that libraries continue to offer important services during the current U.S. economic downturn, such as access to employment materials and educational resources.[3] Libraries continue to be not only early adopters of new technologies, but also early users of cutting-edge technology they see as being effective to their mission of providing information for all. Mobile technology has clearly proven to be a lasting technological innovation that will continue its rise in usage, and not just a fad that will fade away or be replaced by the next innovation. Why then are libraries slow to adopt mobile technology? Most likely the answer to this is the constantly changing, confusing array of mobile devices, operating systems, and supporting technologies. This book will attempt to make mobile technology more understandable and therefore more accessible to librarians, and hopefully encourage adoption of mobile web services and resources as well as enable librarians to better support their increasingly mobile patrons.

DEFINITION OF MOBILE DEVICES

Mobile technology includes all computing devices that allow Internet connection and communication anyplace, anytime. Wireless and cellular technologies allow continued connection to work, friends, and family regardless of our physical location. This technology is becoming ubiquitous in our society today. The benefits of this technology include the ability to be untethered to a workplace or home location and the flexibility to access information immediately at the point of need. It can also be argued that mobile technology increasingly adds to our disconnectedness from other humans and our communities, as can be seen in any modern public place where people are more interested in their mobile devices than the people around them. As with any technology, there are positive effects and negative consequences. Another drawback may be targeted advertising, which is becoming an increasing annoyance on mobile sites such as Facebook. Even vacation retreats are no longer immune to interruptions from work or daily life as cell phone reception approaches worldwide coverage. It will fall to the human users of this technology to regulate it so that it serves us rather than taking over our lives.

Mobile technology is commonly defined as the technology used in cellular communications. It can be more broadly defined as any technology that is portable (battery powered) and uses wireless connectivity. This broad definition would include devices such as cell phones, tablet computers, and even laptops. Pew Research Center reported in September 2013 that 63 percent of American adults who own cell phones go online using their phones.[4] At that time, it was estimated that 91 percent of all adult Americans own a cell phone. More importantly, 34 percent of cell phone Internet users reported that they accessed the Internet mostly through their cell phones rather than with laptops or desktop computers. The survey also reported that the most likely populations to have access to the Internet mostly through their cell phones were young adults, non-whites, and people with low incomes and less education. Clearly, many library users are accessing our services through mobile devices, and we need to plan for providing for their needs.

HISTORY OF CELL PHONES

Cellular phones use radio technology for communication. The term cellular comes from areas or "cells" of radio frequency coverage that were first set up in urban areas and now comprise a worldwide network of towers provided by multiple vendors. When in transit, a cell phone's reception will move from one cell to the next. If it can't find a signal from an authorized provider, it will look for any signal. This is called roaming and usually involves additional charges to the user. Cell phone reception is now available around the world and in some developing countries cellular coverage is more available than wired network infrastructure.

The modern cell phone turned 40 in April 2013.[5] Experiments with the first wireless phones began back in the 1940s and 1950s. These early devices experienced terrible interference and were initially tethered to cars for power. In 1970, the Federal Communications Commission (FCC) finally set aside a radio frequency window specifically for cell phone use. In 1971, AT&T proposed to the FCC the modern system of dividing cities into "cells." The first public cell phone was used in 1973 by Motorola employee Martin Cooper in New York City. This phone weighed 2.5 pounds and the battery only lasted 20 minutes. In 1974, Motorola started selling the Dynatac 8000X to the public for $3,995. This phone was nicknamed "the brick" (figure 1.1) and consisted of a large handset with 20 big buttons, with a long rubber antenna on top. The battery only lasted 30 minutes and it took 10 hours to recharge it.

Since these early days mobile phones have become smaller, cheaper, and smarter. The first "smartphone" was released in 1993. Developed by IBM and called Simon, it was large and bulky like all the existing cell phones of the time, but could access email and send faxes as well as send and receive phone calls. Later in the 1990s, Palm released a much smaller, handheld organizer. This was not a cell phone, but a hand-sized computer with memory and processing power, which was called a personal digital assistant or PDA. It allowed the user to save contacts and keep an address book plus connect to email. Early PDAs also included a calendar, calculator, to-do list, and notepad applications. Later models included games as well as other recreational software. The earliest PDAs used a stylus to enter data before the advent of the touch screen. The small size of this device coupled with the ability to do essential business tasks on the road made it very popular at the time.

The development of these small, handheld devices began even earlier. The first digital pocket-sized organizer was released in 1984 by Psion, the company that later developed the Symbian mobile operating system. Apple introduced the Newton in 1993 and is credited with coining the term PDA. None of these devices were very popular though until the Palm Pilot, which was released in 1996 and introduced a simplified handwriting system for

Figure 1.1. Early "brick" type cell phone

stylus input called Graffiti. Microsoft also got into the PDA market in 1996 with the release of the Windows CE operating system, which could run on multiple hardware devices. In 2001, Microsoft replaced Windows CE with the much more popular Pocket PC, which ran on Dell hardware and became almost as popular as Palm OS devices.

An important feature of the PDA was the ability to save or "sync" information with the user's desktop or laptop computer. At first, a wired cradle that connected the PDA to the desktop computer was needed to sync information, but eventually, wireless technology and cloud applications have made wired syncing obsolete. Many hardware developments such as smaller processors, rechargeable long-life batteries, and wireless networking have contributed to the rise of mobile devices. Technology costs have decreased dramatically. Earlier devices were quite expensive. Today's devices are much more powerful but also cheaper, making mobile technology accessible to many more people than ever before.

By the end of the 1990s and into early 2000, manufacturers began producing devices that looked more like the smartphones of today. These new devices combined cell phone technology with email and Internet access. Modern smartphones, with greatly increased processing power and memory, include global positioning system (GPS) technology and the ability to add applications that extend functionality to rival the computing power previously only available from a desktop computer. The mobilization of computing devices has launched the current mobile revolution, culminating in the meteoric rise in the popularity of the smartphone.

Many types of smartphones are available on the market today. At the end of 2013, the market was split between the Apple iPhone operating system (iOS), which only runs on Apple products, and the Android operating system, which runs on many different types of phones. Market share for the previously popular BlackBerry smartphone has decreased significantly in the past few years. Smartphone vendors are constantly introducing new services and features, trying to corner this vast market. Keeping up with the changes can be a daunting task. Chapter 2 gives more detailed information on specific mobile devices and the supporting technologies.

WHO ARE MOBILE USERS?

Several recent studies of mobile users show that usage of mobile devices is on the rise for almost all populations. Older people and the economically disadvantaged are among the lowest users of technology. In the past international usage of mobile technology, especially in third world countries, has been low but now is also on the rise. This is because of the rapid growth of cellular infrastructure. While wired network infrastructure may barely exist in third world countries, cell phone companies are rapidly providing services and infrastructure worldwide (see figure 1.2). Cell phone usage for access to information on the Internet is, in many developing countries, the most reliable way to access information.

In a 2012 report from the International Telecommunication Union's (ITU) Broadband Commission for Digital Development, the United States ranks only 28th in the world in percentage of people who have access to the Internet.[6] It is estimated that only 70 percent of Americans have access to the

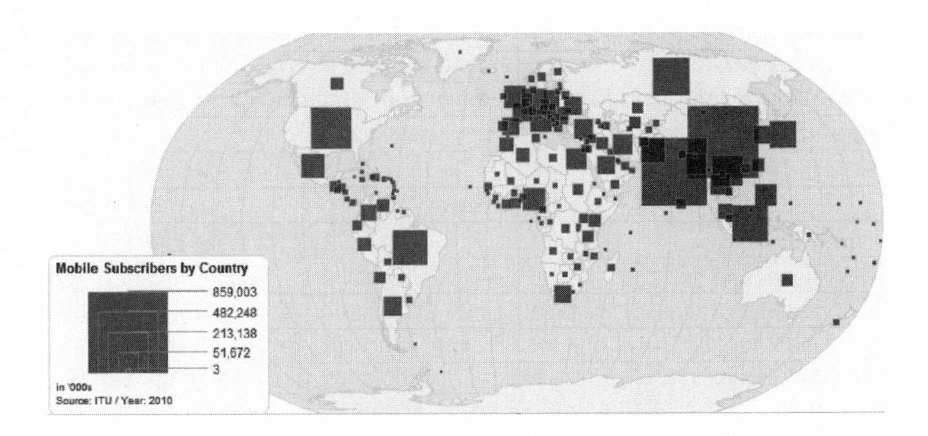

Figure 1.2. Mobile subscribers by country. ***International Telecommunication Union, July 2011***

Internet. The Broadband Commission report predicted that by 2015, other countries such as China and the Philippines will surpass the United States in percentage of Internet users. It is estimated that over 1 billion people worldwide access the Internet using mobile broadband. The growth in international access to the Internet is not surprising due to the growth of the global online economy. Internet access also greatly increases people's access to education and jobs. A major goal of the United Nations is for 60 percent of the world to have Internet access by 2015, with 50 percent of the access occurring in developing countries.

The ITU Broadband Commission report predicts that with the growth of wireless broadband infrastructure worldwide, the digital divide may no longer be between those who have Internet access and those who don't, but between those who have better access through greater speed and functionality and those who have limited access. According to the report, some trends to look for in the future include: better access for all to health care, government services, and jobs; collaborative "crowd-sourcing" of publishing and news reporting; worldwide project management; and multicountry, collaborative disaster relief efforts. Storage of data in the cloud means less reliance on physical hardware, but security and privacy issues will be concerns.

In December 2012, Pew Internet Research Company issued a report entitled "Mobile Connections to Libraries."[7] Pew's survey of Americans 16 and older found that 13 percent of respondents have accessed library resources from mobile devices. Top users include women, people with at least some college education, and parents with minor children. Use is highest in the age groups 18–29 and 30–49. For all Internet users, the most commonly accessed items from library web pages include listings of library hours, catalog search interfaces, book reservation and renewal, online database links, and information on library programs or events. With a growing number of people accessing library websites from mobile devices, it is imperative that library web developers create a mobile strategy and at least provide a mobile-friendly layout. More information on providing mobile content from your library website will be provided in chapter 3.

MOBILE APPS

There is an app for every need. Resource vendors, start-up companies, and existing corporations all are tapping into the vast growth in mobile technology by creating and marketing mobile apps optimized for use on smartphones and tablet devices. It is getting easier to create apps, but keeping up with the rapidly changing technology complicates the process. Apps created for the Apple iPhone will not work on Android devices, and neither will work on other operating systems. Some vendors and businesses are opting to mobilize websites rather than create apps, in order to avoid having to create multiple versions for different devices. The ease of use and popularity of apps is important, though, when considering what format mobile access to your website will take.

Cloud technology goes hand in hand with mobile technology. Cloud services are technical services that are provided via the Internet. Services that traditionally were provided by in-house technology departments are now being replaced by third-party vendors on the web to avoid the cost of hardware, software, and maintenance, as well as personnel costs. These services might include document sharing, platforms for online resource sharing, web conferencing, social networking, and off-site storage of data. The risks of switching from locally hosted and controlled information technology to cloud-based resources are often outweighed by reduced costs and access to better services. Security issues and the risk of losing data or investing in a company that ends up going out of business are concerns, and should be weighed against costs savings and ease of use. Many cloud services provide a mobile app for easy access to their services and resources. See chapters 4 and 6 for more information on mobile apps, such as how they are developed and which apps can be of use in libraries.

MOBILE TECHNOLOGY IN THE LIBRARY

Because increasing numbers of patrons use mobile technology, all types of libraries must consider what needs to be done to support the technologies their patrons use. Many library users are now also primarily mobile users, so a library risks losing this group of users if it does not provide mobile access to its collections and services. Many patrons are becoming more comfortable using mobile devices for information seeking, and prefer the quick access to information that mobile devices provide. Library resource vendors are increasingly aware of this fact and are providing mobile apps with access to their information. In order to stay relevant in today's mobile information environment, librarians should also become well-versed in providing mobile reference services via text or chat messaging.

Librarians can also benefit from using mobile devices when providing outreach and communication with users beyond the library. Presentations and teaching can be facilitated by mobile devices. Productivity apps for note taking, document sharing, and research should become familiar to all librarians, especially in academic settings. The built-in GPS capability along with new mobile technologies such as QR codes should be explored for creative use in libraries. Libraries and librarians need to keep up with the constantly changing world of mobile technology or risk becoming obsolete in the increasingly mobilized future. Mobile technology can be used to provide new, innovative services to our users, who may think of libraries as just places for physical books. Libraries and librarians as providers of information can remain relevant to the future by embracing new technologies and finding better ways to reach out to our users with innovative information services. More information and details on providing innovative library services using mobile technology can be found in later chapters of this book.

In planning for introduction of mobile services in your library, a mobile strategy is extremely important. This strategy may include mobilizing your website, adding mobile services, and providing outreach via mobile applications. Mobile strategies, as well as many kinds of mobile services that libraries can provide, will be explored further in later chapters of this book. When considering all of the many mobile services that can be provided by your library, a mobile strategy will help you prioritize which services may be the most important for your library and your users. Questionnaires and focus groups may also be helpful in choosing what services to deploy first in your library. Careful planning and assessment will assure the success of your library's mobile initiatives, and will be especially important with tight budgets restricting what projects can and cannot be funded. Best practices for libraries implementing mobile services and websites will be covered in chapter 7.

FUTURE DIRECTIONS

What will the future of mobile technology bring? According to *Time* magazine, we are entering the golden age of mobile technology.[8] Infrastructure and small-scale technology solutions are well established and in place to support the next generation of mobile innovations. Bendable plastic displays, longer-life batteries, built-in GPS technology, and data storage advancements have made new and exciting innovations possible. Some trends to look for include "intelligent" apps based on saved personal data, mobile health care apps, and mobile apps that anticipate your needs such as automatic booking of flight reservations and forwarding traffic alerts. In the future, your smartphone may be your "digital wallet," taking care of all your banking and purchasing needs.

The next big trend in mobile technology may be wearable devices. I recently saw a person at Starbucks wearing a Google Glass, which at this writing is only available to a select few beta testers. Google Glass is eyewear with a built-in mobile device that allows the wearer to take pictures and video, get directions, send messages, and get information by using voice commands. Another innovative company, Pebble, has now released a "smart" watch and will soon open an app store for these devices. The Pebble watch displays email, text and phone notifications, and has apps for fitness, music, and games. Virtual reality and gaming applications for mobile devices are also predicted to become more common in education and business applications as well as entertainment. Will your library be ready for these innovations? Chapter 10 will discuss each of these trends in more detail and give ideas and resources for keeping up with mobile technology innovations and anticipating future needs and directions.

NOTES

1. Cruz, Laurence. "Mobile Internet Tops 12 Most Disruptive Technologies." The Network: Cisco's Technology News Site, September 2, 2013. http://newsroom.cisco.com/feature-content;jsessionid=CF092F3F5F1C867F543D8B384C357D59?type=webcontent&articleId=1255430 (accessed February 26, 2014).
2. Duggan, Maeve, and Aaron Smith. "Cell Internet Use 2013." Pew Research Centers Internet American Life Project RSS, September 16, 2013. http://www.pewinternet.org/2013/09/16/cell-internet-use-2013/ (accessed February 26, 2014).
3. "State of America's Libraries Report 2013." American Library Association, 2013. http://www.ala.org/news/state-americas-libraries-report-2013 (accessed February 26, 2014).
4. "Mobile Connections to Libraries." Pew Internet Libraries RSS, December 12, 2102. http://libraries.pewinternet.org/2012/12/31/mobile-connections-to-libraries/ (accessed February 26, 2014).
5. "Cell-ebration! 40 Years of Cellphone History." Mashable, April 3, 2013. http://mashable.com/2013/04/03/anniversary-of-cellphone/ (accessed February 24, 2014).
6. "The State of Broadband 2012: Achieving Digital Inclusion for All." Broadband Commission, September 2012. http://www.broadbandcommission.org/Documents/bb-annualreport2012.pdf (accessed February 26, 2014).
7. "Mobile Connections to Libraries." Pew Internet Libraries RSS, December 31, 2012. http://libraries.pewinternet.org/2012/12/31/mobile-connections-to-libraries/ (accessed February 26, 2014).
8. Bajarin, Tim. "Welcome to the Golden Age of Mobile." *Time*, August 26, 2013. http://techland.time.com/2013/08/26/welcome-to-the-golden-age-of-mobile/ (accessed February 27, 2014).

FURTHER READING

Cuddy, Colleen. *Using PDAs in Libraries: A How-To-Do-It Manual*. New York: Neal-Schuman Publishers, 2005.

"Gartner Says Smartphone Sales Grew 46.5 Percent in Second Quarter of 2013 and Exceeded Feature Phone Sales for First Time." Gartner, August 14, 2013. http://www.gartner.com/newsroom/id/2573415 (accessed February 27, 2014).

Rosen, Christine. "The New Meaning of Mobility." *The New Atlantis*, Spring 2011. http://www.thenewatlantis.com/publications/the-new-meaning-of-mobility (accessed February 27, 2014).

Shwayder, Maya. "One-Third of World's Population Using Internet, Developing Nations Showing Biggest Gains." *International Business Times*, September 24, 2012. http://www.ibtimes.com/one-third-worlds-population-using-internet-developing-nations-showing-biggest-gains-795299 (accessed February 27, 2014).

RESOURCES

A Brief History of Smartphones: http://www.techhive.com/article/199243/a_brief_history_of_smartphones.html
Cell Phone Timeline: http://iml.jou.ufl.edu/projects/fall04/keith/history1.htm
The Future of Mobile Technology: http://visual.ly/future-mobile-technology
History of the Smartphone: http://www.qrcodescanning.com/smartphonehist.html
Pebble: https://getpebble.com/
Project Glass Could Be Called Google Eye When It Lands on Your Face: http://techcrunch.com/2012/04/05/google-eye/
What You Should Know about Flexible Displays (FAQ): http://news.cnet.com/8301-1035_3-57607171-94/what-you-should-know-about-flexible-displays-faq/

Chapter Two

Overview of Mobile Devices

There are many types and models of mobile devices available today. These devices can be divided into two main types: smartphones and tablets. Smartphones make up the majority of devices available today. There are many models to choose from, and manufacturers are constantly coming up with new features in order to market new devices. In this chapter we will give an overview of the most popular models and features available, as well as discuss topics of concern to those thinking of implementing library mobile programs.

SMARTPHONES

As mentioned in the previous chapter on the history of mobile devices, smartphones were preceded by PDAs, some of which are still in use today. PDAs were the first portable, handheld devices that had some of the features of full-sized computers. PDAs kept track of calendars, notes, and contact lists and could access email and documents when connected to a desktop computer. Later models also included games and other add-on programs, making them more like desktop computers, but they were limited by the amount of memory and processing power available at the time. Advancements in hardware technology, making electronic components smaller and cheaper, made the concept of a truly "smart" phone a reality.

While cellular phones have been around for over 40 years, the first true smartphone was manufactured by IBM in the early 1990s. This was a very large bricklike device barely resembling anything in use today. Not until 1998 was a more modern-looking model released, the Nokia 9110 Communicator, which was closer to the size of phones today and featured a flip-out keyboard. The Communicator screen display was still black and white, and the device could not connect to the Internet.

In 2002 Research in Motion Limited (RIM) developed the BlackBerry 5810. This was the first device that could download email and surf the web. The first model required headphones to speak on the phone, but later models became headset free. BlackBerry devices became especially popular in business settings, and RIM sold many models to busy executives who were almost addicted to these devices because they could continue to work while outside the office and send emails directly from the phone without dialing in to a server. RIM also provided their own wireless connectivity, which was available as a subscription service to BlackBerry users only. Today, BlackBerry is still in use, mostly in businesses, but market share greatly decreased after the more popular Android and Apple consumer devices were developed. Microsoft also recently entered the field of smartphone sales with its Windows phone, which has been growing in popularity, although lagging far behind Android and Apple.

Most smartphones today use touch screen technology and surpass the computing power of early computers. Smartphones can be classified by the software that runs on the device, much like desktop computers can be classified as Windows, Apple, and Linux workstations. Figures 2.1 and 2.2 show the top smartphone operating systems in use today and the world market share for each. The Android operating system is open source like Linux and runs on phones manufactured by many different vendors, and therefore has the biggest global market share of smartphone sales. Apple mobile devices are very popular but lag behind Android in market share due to the limited number of devices sold using their proprietary iOS. In the following sections, each of the most popular smartphone operating systems are discussed in more detail.

Android

Almost 80 percent of smartphones in use today are Android phones.[1] Android is an open-source operating system owned by Google and based on Linux. The first Android smartphone was released in 2008, and by fall 2013 over 1 billion had been sold worldwide. Because the Android operating system is open source, multiple phone vendors manufacture Android-based devices. Samsung is the most popular smartphone vendor today with 24.7 percent of the market share in the fall of 2013. Nokia and LG Electronics are also popular with 14 percent and 3.9 percent of the market share respectively. Free and paid apps are available for Android phones through Google Play, Amazon, and other sources. By summer 2013 there were over a million apps available, and the numbers are constantly increasing. Android phones use the Google Chrome browser to access the Internet, and come with Google Maps,

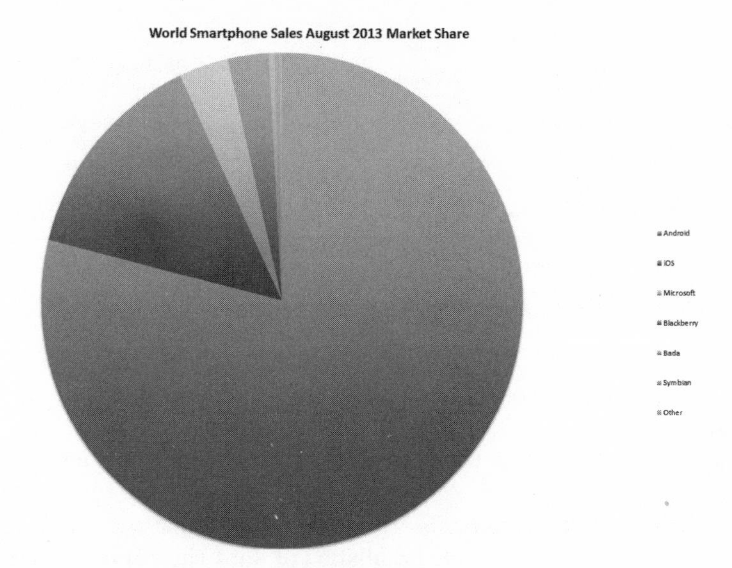

Figure 2.1. World smartphone sales. ***Gartner***

Gmail, and Google Drive apps. Android has a built-in voice command feature that can be used to activate some of the phone functions.

The first phone running the Android operating system was manufactured by HTC. Motorola decided to adopt Android as the operating system for all their phones in 2009, and then released the extremely popular Droid smartphone. Sales of Android devices began to take off as more vendors adopted the open-source operating system for their devices. It is estimated that by 2010 almost 200,000 Android phones were being activated per day.[2] The advantage of open-source software is that any programmer can modify the code and therefore it can be modified to run on any hardware, making it available to any phone vendor. This encourages competition among vendors and also encourages innovation as manufacturers come up with new features and exciting new technology that will appeal to the consumer.

Apple iOS

Apple released the first iPhone in 2007 and *Time* magazine voted it the invention of the year. This was one of the first mobile phones to use touch screen technology, which incorporated tap, pinch, and zoom gestures in order to improve the browser experience on a small-screen device. The keyboard function is provided through a software interface. Previous mobile devices used a stylus or keyboard for input. The iPhone was extremely popular with consumers because it integrated music, video, phone, and Internet in a simple

OS	Type	Apps Available	Development	Owner
Android	Open source	Yes	Active	Google
iOS	Proprietary	Yes	Active	Apple
Microsoft	Proprietary	Limited	Active	Microsoft
Blackberry	Proprietary	Limited	Declining	RIM/Blackberry
Symbian	Open source	Limited	Inactive	Symbian Ltd

Figure 2.2. Major cell phone operating systems

but powerful user interface. Unlike Android, Apple's iOS operating system is proprietary and only runs on Apple devices. Originally only running on the iPhone, iOS was later adapted for use on the iPod Touch and the iPad. The original iPhone did not allow the installation of third-party apps, but with the release of the iPhone 2, Apple also announced the opening of the App Store, which quickly became very popular. Apps allow access to Internet information at the touch of a finger without having to use a browser. By 2011, 15 billion total apps had been sold and there are now over 1 million unique apps in the App Store.

Apple's innovative and user-friendly iPhone design made it initially the top-selling smartphone operating system, but it was surpassed by Android in U.S. sales in 2010. Due to the propriety nature of the operating system, Apple has fallen behind Android in market share. There are still many more apps available for iOS than for Android, but the gap between them is closing. The iPhone uses the Safari browser and comes with iTunes and Apple Maps, which features traffic information, 3D views and turn-by-turn navigation. The newer models have the built-in voice command feature, called Siri, which allows you to activate phone features using only your voice.

Microsoft

Microsoft was a late entry to the smartphone market, releasing its Windows Phone in 2010. Prior to Windows Phone, Microsoft marketed the Windows mobile operating system, which ran on the Pocket PC PDA. Microsoft partners with multiple phone vendors such as LG, Samsung, Nokia, and HTC to run its mobile platform. The Windows Phone runs limited mobile versions of most Microsoft applications such as Word, Excel, and PowerPoint plus integrates with other popular services such as Facebook, Twitter, and Google accounts. Microsoft's Windows Phone features the Internet Explorer browser and it uses the Windows 8 interface, which allows you to "pin" favorite apps to the desktop. It also features live updating of apps. At the end of 2012, there were just over 100,000 apps available for the Windows Phone. Gaining in popularity, the Windows Phone has now surpassed the BlackBerry but is still far behind Android and Apple iOS in market share.

BlackBerry

RIM released the first BlackBerry device in 1999, which was essentially a pager with email capability. Since then, there have been many BlackBerry models released. The first BlackBerry models only had keyboard input, but newer models have added touch screen interfaces. BlackBerry devices focus on email applications but also include a camera and can play music. The BlackBerry has a built-in propriety web browser, which has been significantly upgraded in version 10. A limited number of apps are also available. In 2012, there were 80 million BlackBerry subscribers worldwide. With the rising popularity of the Android operating system, BlackBerry market share has significantly decreased.

BlackBerry was the first mobile device released that provided constant wireless access to email. It also had its own text messaging system that allowed text communication between BlackBerry users and it could sync information directly with the user's desktop computer. BlackBerry focused on providing a mobile solution for business applications and was eclipsed by the release of the consumer focused Apple iPhone with seamless multimedia integration and user-friendly design.

Symbian and Bada

The Symbian mobile operating system was originally developed by Psion and is an open-source platform like Android. In 2002, Nokia began marketing smartphones running the Symbian operating system. It quickly became the most popular operating system for many other phones including Samsung, Motorola, and Sony models in addition to Nokia. When Nokia switched to using the Windows Phone operating system in 2011, Symbian lost most of its market share almost overnight. Symbian-based phones were the earliest to branch out from a purely business focus to include video, music, and gaming. Because of this early focus on entertainment, the Symbian operating system was a significant milestone in the development of modern smartphone technology. Samsung was recently developing a new operating system called Bada, which has a small market share. In early 2013, Samsung stopped active development of Bada, moving to a new proprietary operating system called Tizen, which is Linux based.

TABLET COMPUTERS

Tablet computing devices of today are similar to smartphones but with larger displays and more computing power. Tablet-like devices have been around for many years, but not until the smaller, lightweight models of today were released did the concept truly become popular. Touch screen capability, built-in wireless, and portability make tablets a desirable alternative to carrying a heavy laptop. With a large number of apps available for a variety of information needs, tablet computers are becoming a viable replacement for laptops. Cloud applications allow data to be stored remotely and synchronized between devices. In many daily uses, there is no longer a need for the powerful processor and large amounts of file storage available with a desktop or laptop computer. Tablet computer sales rose by over 50 percent in 2013 while sales of traditional PCs declined.[3] Continued exponential growth seems assured. In a 2013 TechCrunch article, Gartner is quoted as predicting tablet sales to surpass PC sales by 2017.[4]

Apple

Apple's extremely popular tablet computers, the iPad and the new iPad mini, are the top-selling tablet computers today. The original iPad was released in 2010 with options for wi-fi only or wi-fi and 3G service available from AT& T. The iPad 2 was released in 2011 with a faster processor. In 2012, the fourth-generation iPad was released as well as a smaller 7.9 inch version, the iPad mini. In 2013, Apple released the lighter and thinner iPad Air and a new version of the iPad mini with an upgraded display. Cellular data plans are now available from Verizon, Sprint, and T-Mobile as well as AT&T. The iPad introduced gyroscope technology, which allows the display to rotate as the device is turned. It also introduced pinch and zoom gestures that allow text and graphics to be resized with touch technology.

Android

Samsung has the next best-selling tablet computer after the Apple iPad with its Galaxy Tab series. The Galaxy Tab comes in 10.1 inch, 8.9 inch, and 7 inch sizes. Samsung uses the Android operating system and apps are sold through Google Play. Improvements have been steady through the release of the current third-generation device, but sales have lagged behind Apple. Both wi-fi and cellular options are available. Cellular plans are available from all the major vendors. A major selling feature of the Samsung Galaxy series is the multitasking feature and multitouch technology. Samsung also markets the Galaxy Note, which uses stylus input for written note-taking.

Other Tablets

Other tablet computers include Amazon's Kindle Fire, Google's Nexus, Microsoft's Surface, and the Asus MeMO. The Kindle Fire is a version of Amazon's popular Kindle e-reader but with full color and access to videos, games, and apps as well as the Internet. Google released the Nexus 7 inch and 10 inch tablets in 2012. With access to apps from Google Play, the popularity of the Nexus is rising although it is relatively new to the tablet market. Also in 2012, Microsoft released the Surface tablet computer. It comes with a unique touch cover keyboard as an accessory and features a scaled-down version of Microsoft Office. The newest version, the Surface Pro 2, comes with the Windows 8.1 operating system. Sales of the Surface tablet have not been as high as expected. Another new entry to the tablet market is from hardware manufacturer Asus. The Asus MeMO tablet is a newer version of the Android-based Nexus and is relatively inexpensive. As new models of tablets come out and the technology improves, these handheld devices are also beginning to approach the power of laptop computers.

E-READERS

A subset of tablet computers, e-readers are created especially for reading e-books. These e-book readers are limited to e-text content and also usually provide basic wi-fi connectivity to the Internet. The e-reader screen provides better readability than a tablet, especially in bright light, and battery life is much longer. Popular e-book reader models include Amazon's Kindle, Barnes and Noble's Nook, the Kobo, and Sony Reader. Each model supports different file formats in addition to their own proprietary formats. Early Kindle models only supported the Mobipocket file format, but later models have added support for PDF, ePub, HTML, and other formats. Many e-reader models do not support graphics display, but newer models such as the Kindle Fire are closer to a tablet computer with color and graphical displays. As tablets become more popular, they are displacing e-readers in the marketplace since they can display e-books as well as provide much more functionality.[5] These e-book readers are typically cheaper than tablets and are reportedly more popular with older population demographics.

SUPPORTING TECHNOLOGIES

Mobile devices would not be possible without the development of several supporting technologies including wireless communications. There are three major types of wireless connectivity in use on mobile devices today: wi-fi, cellular, and Bluetooth. Smartphones typically are cellular devices, but modern models also have wi-fi built in. Tablets can be wi-fi only, but many models are also available with built-in cellular connectivity. Most mobile devices also have Bluetooth capability for connecting wireless accessories such as keyboards and mice. In the following section, we will discuss these technologies in more detail.

Wi-fi connectivity has been around for many years and became popular in the early 2000s. Wi-fi is the wireless equivalent of an Ethernet (wired) local area network. This technology is based on radio frequency and employs a transmitter and receiver. Wi-fi connectivity relies on wireless access point hardware that transmits a wireless signal that can be picked up by an antenna located in mobile devices. This model is considered a local area network because devices must be within a hundred feet or so of the wireless access point in order to connect. Wi-fi networks are located in many academic institutions, businesses, airports, malls, coffee shops, and now even in some outdoor areas of major cities.

Wi-fi is based on the IEEE standard of 802.11. The 802.11b standard was the first popular standard and had a data transfer speed of up to 11 megabits per second (Mbps). Later standards were developed to be much faster, such as 802.11g and 802.11n. In 2012, the 802.11ac standard was released with a maximum speed of 433 Mbps. The speed of the wi-fi connection is also dependent on how close the device is located to the access point as well as how many people are connected to the access point at the same time. Multiple access points can be deployed in a high-use area, and switching between points is now automatic. Wi-fi networks can be open or protected with a wireless encryption method such as WEP or the stronger WPA standard, both of which require a password in order to connect.

Cellular technology is based on voice transmission technology. The first generation of phones used analog voice technology, and we entered the second generation with digital voice systems in the 1990s. Modern cell phones use the much faster third-generation (3G) and fourth-generation (4G) technologies. There are currently several different versions of 4G cellular, with LTE being the most recent and fastest version. New and faster cellular technology is being developed all the time, but is not necessarily available in all locations. Some cellular plan carriers have better coverage and are faster at implementing improved technology.

Networking of cellular devices is made possible by wireless mobile broadband technology, which delivers access to the Internet from towers owned by mobile service providers or "carriers." Mobile broadband connections are available wherever mobile service carrier company towers are located. These towers transmit over long distances, and switching between carrier towers is automatic. Theoretically, mobile coverage is located almost anywhere in the United States and most other countries, but the coverage varies by company. Mobile network coverage is also increasing rapidly in third world countries that may be lacking in wired network infrastructure, making wireless broadband technology much more important in these countries.

There are many mobile network carriers, which also vary from country to country. Most carriers require monthly or yearly data plans. Fees are based on data usage. Unlimited plans are available, but more expensive. Some mobile companies have better coverage than others, but most allow "roaming" access where mobile devices can connect to another carrier's tower for an added fee. Different countries also have different carriers, and roaming fees can get expensive when crossing international borders. The European Union has regulated roaming fees between European countries, and other areas are also exploring international roaming agreements. In the future, cellular coverage may be ubiquitous all over the world.

Cell phones contain portable wireless modems, which allow them to connect to mobile carrier towers. Early modems were large and slow, but over the years they have become smaller and faster, which makes today's smartphone technology possible. A 4G mobile connection to the Internet is much faster than a wi-fi connection, although wi-fi is typically more reliable. Wi-fi is only available near a wireless access point, which is connected to a wired network. Cellular mobile access allows more freedom to access the Internet from anywhere, but there are gaps in coverage where a tower signal may not be available. Mobile broadband cards can also be purchased and installed in laptops to connect to mobile towers, making even laptops portable beyond the reach of wi-fi networks.

Cellular plans for mobile devices were originally limited to specific carriers per device, but now most devices with 3G or 4G built in give the user the option to purchase a cellular plan from any of the top vendors. Since smartphones and tablets access the Internet and download data, cellular plans that allow for large enough data usage should be purchased to keep costs to a minimum. An unlimited data plan is optimal especially for tablets. Of course, since wi-fi is also built in, there will be no charges when connected to wi-fi access points. Apple devices allow the user to turn off cellular data traffic to control charges. Android devices also have settings that can help minimize data usage. Data charges apply whenever the Internet is accessed, but also when apps push notifications to your device. Some apps load data in the background, which also increases data usage. Android devices allow you to check which apps are using too much data and to restrict the use of background data. Apps that use the cloud for storage employ data charges when they sync from your device to the cloud. You should consider updating apps only when connected to a wireless network, in order to save cellular data charges.

There are a wide variety of accessories that can be purchased with mobile devices. Cases and screen covers are available for every model. Cases for tablets that contain keyboards that connect wirelessly using Bluetooth are available. Be aware that with cases, one size does not fit all; the user must purchase the correct case for the model of mobile device. Mobile devices also come with a variety of cables. The power cables usually connect to USB so that the user can charge them from a computer and the USB cable can plug into a power adapter for direct wall socket power charging. Power cables are unique to devices and are not always interchangeable. Cables for displaying from mobile devices are also available. These allow connection to a monitor, TV, projector, or other display device. Unfortunately, as technology changes, the connectors change too and are not interchangeable, requiring all new cables for each new device. The user needs to know whether the connection is to VGA, DVI, or HDMI displays as well as the type and size of the mobile device's output port. Apple is particularly notorious for changing the type of connection ports between models.

Specialized accessories are also beginning to appear for mobile devices. Fitness accessories such as heart rate monitors are popular. Wireless speakers are available as well as other recreational devices such as a golf swing analyzer. Health monitoring devices are now being manufactured, connecting and recording information on smartphones, such as blood pressure and glucose monitors (figure 2.3). There is also a smart scale that sends weight information to the user's phone and stores it over time, and even a breathalyzer device that connects to the iPhone via Bluetooth to determine whether or not it's safe to drive.

Printing from mobile devices is still under development. Newer printers are equipped with wireless, but successfully connecting from a mobile device can be difficult. HP printers come with ePrint software to facilitate wireless printing. Apple devices are equipped with AirPrint that works with many printer brands. For Android devices, Google's CloudPrint works from the Chrome browser or Google apps. Of course, documents can be saved using any of a number of cloud apps and then the document opened later on a computer or laptop to print. The ability to take screenshots is built in to most devices. On an Apple device, the user can press the home and on/off button at the same time and a screenshot will be sent to a camera roll. The process for taking screenshots on an Android phone differs between models. Newer models take screenshots by pressing the volume down and power buttons at the same time. The screenshot will be sent to the Gallery app. Some models work by pressing the power and home buttons, but some older models need add-on software to be able to do screenshots.

Mobile devices are primarily single-user devices. They are connected to a personal user account and are highly customizable for user preferences. It is difficult to share mobile devices between users, for example in a business environment, because of personal settings on the device. Although tablets are usually attached to a single-user account, it is possible to create a standard profile template, which can be copied to multiple tablet devices. This is very

Figure 2.3. Withings smart blood pressure monitor. *www.withings.com*

helpful in a library setting where tablets might be checked out to patrons. Standard apps and files can be installed on the tablet from a template, and the tablet can be reimaged quickly if patrons download unwanted content or delete any of the standard apps. More about imaging tablets will be discussed in chapter 4.

A WORD ABOUT SECURITY

Because smartphones and tablets are essentially the same as desktop and laptop computers, they are also vulnerable to some of the same security risks. There are software vulnerabilities and browser threats as well as email and text messaging threats. Android devices are especially vulnerable due to the open-source nature of the operating systems. There are apps now available to protect mobile devices from security threats such as malware and viruses, but there are also simple safety measures that users can follow to help keep devices and data more secure. One of the most important security measures for any device is to keep the software updated. As new threats come out, operating system updates are developed to plug security holes. Another simple way to keep mobile devices safe is to lock the device with a password or pin. The iPhone 5s comes with a fingerprint scanner, which makes this device the most secure on the market today. With these measures, the saved data on a device will not be readily available if it is lost or stolen.

Another way to help keep a device secure is to never connect to banking or other secure websites from public wi-fi connections where passwords may be compromised. Use the same precautions with email and messaging that you would use on your personal computer or laptop. Never open a suspicious email attachment or download something that came from a suspicious source. There are viruses that can hijack a smartphone and even damage or shut down the device. Spyware can be accidentally installed, and can compromise passwords by recording and transmitting information. Apple iCloud has a feature that allows users to track missing devices remotely by signing in to an account from any computer. The Find my iPhone app locates a device on a map. In iOS 7, a user can activate a locking feature as soon as a device goes missing so that no one can access it without the user's login information. Users can also initiate a remote wipe if they suspect the device has been stolen so that the device is reset to factory settings and all personal data is removed. Following simple security steps will keep devices and data safer and free from harmful malware.

In the next chapter, we will discuss making websites more accessible to mobile users. There are many ways to make web content more compatible with the many types of mobile devices that may be accessing a website. We will explore several options and provide guidelines for making websites more mobile-friendly.

NOTES

1. Tofel, Kevin. "Android Sales Overtake iPhone in the U.S.—Tech News and Analysis." Gigaom, August 8, 2010. http://gigaom.com/2010/08/02/android-sales-overtake-iphone-in-the-u-s/ (accessed February 27, 2014).

2. Ahmad, Majeed. *Smartphone: Mobile Revolution at the Crossroads of Communications, Computing, and Consumer Electronics*. North Charleston, SC: CreateSpace, 2011.

3. Lomas, Natasha. "Tablets to Grow 53.4% This Year, Says Gartner, As the Traditional PC declines 11.2% [Updated]." TechCrunch, October 21, 2013. http://techcrunch.com/2013/10/21/tablets-vs-pcs/ (accessed February 27, 2014).

4. Lomas, "Tablets to Grow 53.4% This Year."

5. Gross, Doug. "As Tablets Boom, e-readers Feel the Blast." CNN.com, February 28, 2013. http://www.cnn.com/2013/02/28/tech/gaming-gadgets/tablets-replacing-e-readers/ (accessed February 27, 2014).

FURTHER READING

Cha, Bonnie. "House Call: Five Smartphone Accessories That Help Monitor Your Health." *AllThingsD*. July 3, 2013. http://allthingsd.com/20130703/house-call-five-smartphone-accessories-that-help-monitor-your-health/ (accessed March 1, 2013).

Schneiderman, Ron. *The Mobile Technology Question and Answer Book: A Survival Guide for Business Managers*. New York: AMACOM, 2002.

RESOURCES

11 Ways to Trick Android into Using Less Data: http://howto.cnet.com/8301-11310_39-57599576-285/11-ways-to-trick-android-into-using-less-data/

Cellular Generations Definition, from PC Magazine Encyclopedia: http://www.pcmag.com/encyclopedia/term/55406/cellular-generations

Cellular vs. Wi-fi Definition, from PC Magazine Encyclopedia: http://www.pcmag.com/encyclopedia/term/57165/cellular-vs-wi-fi

iPhone and iPad (Cellular Models): Understanding Cellular Data Settings and Usage: http://support.apple.com/kb/ht4146

Two Ways to Hit "Print" on a Mobile Device: http://allthingsd.com/20130709/two-ways-to-hit-print-on-a-mobile-device/

Wireless Minneapolis: http://www.minneapolismn.gov/wireless/index.htm

Chapter Three

Mobilizing Your Website

With the rise of mobile devices, more and more people are accessing websites from their phones and tablet computers. In March 2013 *EdTech* magazine reported that 45 percent of all adults owned a smartphone and that in 2013, mobile devices would surpass the world's population.[1] In December 2012 the Pew Research Center reported that 13 percent of people ages 16 and older had accessed library websites from mobile devices.[2] It's easy and fast to check library hours, look up a book, or check your library account right from your mobile phone at the time you think about it rather than waiting until you have time to sit down at a desktop computer.

In its 2010 Mobile Libraries Survey, *Library Journal* reported that almost 40 percent of libraries responding had no mobile services.[3] Obstacles to implementing mobile services included conflicting priorities, insufficient budgets, and insufficient skills to support these services. Only a couple of years after this survey, there has been an explosion of tools for developing mobile services and most libraries are at least considering mobilizing their library website in some way. Many tools are free and are getting easier to use, requiring little technical background for an entry-level solution. This chapter will primarily address ideas for mobilizing your library's online web presence. Chapter 4 will provide information on the more technical and expensive solution of creating mobile apps.

So how do you get started building your library's mobile presence? This chapter will guide you through some of the options, best practices, tips, and tools. There are many options for creating a mobile web presence. It can be difficult to know where to start and what solution is right for your library. The first step in mobilizing your library website is to thoroughly understand your users and their needs. Your library will be unique in many respects from other libraries, so it is crucial to design for actual user needs so that your mobile project is successfully implemented.

PREPARING FOR A MOBILE WEB PRESENCE

Web statistics programs such as Google Analytics can provide useful information about how many people access your website from mobile devices, what devices are used, and what pages are most commonly accessed by mobile devices. Google Analytics, which is free for a basic account, has a section on statistics about the mobile audience, which includes an overview of mobile visits and the devices used. There is also a Google Analytics for Mobile Websites product that will collect full visit and traffic stats for mobile websites when installed on the server. Other popular web statistics programs include WebTrends, Stat Counter, WebSTAT, and Open Web Analytics.

During a one-month sample in the fall of 2013, about 2 percent of the University of Washington Health Sciences Library website visits were from tablet computers and another 2 percent from other mobile devices (see figure 3.1). There were over 5,000 mobile visits for the month studied. This is over twice the number from a year ago. The Apple iPad accounted for about 55 percent of the devices and the Apple iPhone accounted for another 27 percent. Android phones made up most of the rest of the devices accessing the website. This site has a simple mobile style sheet but is not fully optimized for mobile. With the current rate of mobile device usage, it is definitely time to increase web optimization on this library website.

Especially for academic libraries whose next-generation students have grown up with mobile devices, the need to start considering what their websites look like from smaller screens is pressing. Public libraries too may be missing out on a whole segment of their user population by ignoring mobile device users, especially for youth programs. Library websites that serve international researchers need to know that many developing countries have better access to cellular networks than wired infrastructure and may be accessing resources from phones.

Medical libraries with a clinical focus have an opportunity to work with physicians and other health professionals who are excited about the portability of mobile devices and the ability to use mobile-ready clinical tools at the bedside. There are many new apps released each year specifically for the medical professional or health market. We will take a more in-depth look at some of these apps in subsequent chapters.

A good way to identify a library's potential web audience is through technology usage surveys. These can be conducted along with your library's annual survey activities by adding a few questions about device usage and mobile habits. An in-library survey could also be conducted by asking library

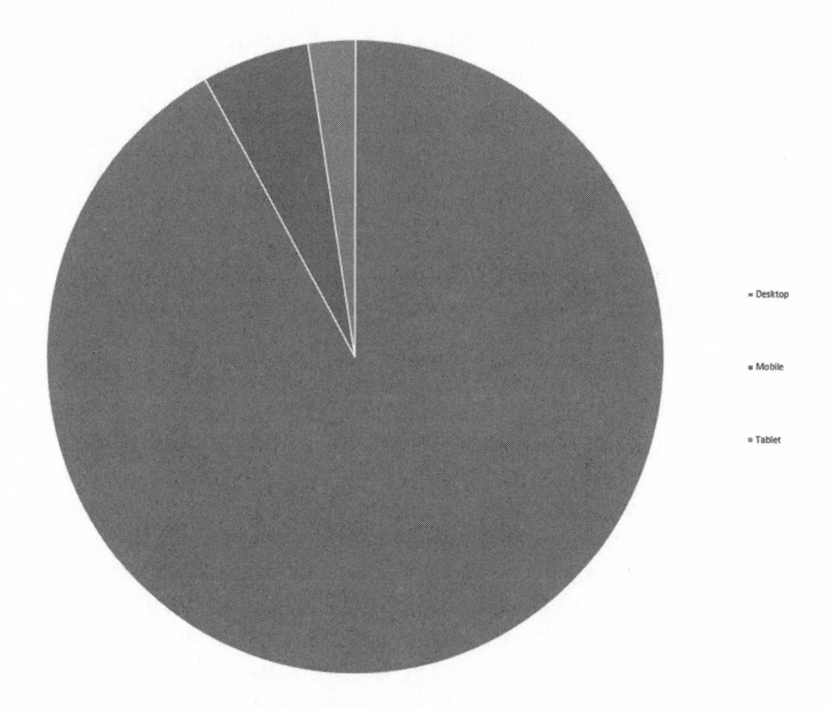

Figure 3.1. Sample usage chart from Google Analytics

visitors to complete a short questionnaire about your website and their mobile needs.

CREATING A MOBILE STRATEGY

Realizing that reaching a growing mobile audience is increasingly important, libraries need to develop a formal strategy to guide their steps toward reaching this important audience. The first step in any web design project is to thoroughly know your audience. This step is also crucial for creating a mobile strategy. We have already explored the usage of web analytics to get a picture of what devices connect to our website and what percentage of users are connecting from mobile devices. During the design process for your library website, most likely a document was created that outlined your library's key audiences and needs. Usability studies may have been done on the library website that will also provide a good idea of your patron's needs. When developing a mobile presence, it is extremely important to be very clear about your website users and their needs before even beginning the project.

Developing a formal mobile web strategy gives a guiding direction to your mobile project. If done right, it will help ensure a successful mobile presence for your library. There are several key ideas that should optimally be addressed in a mobile web strategy statement. The first and foremost thing to consider is the "why" of the project. Consider your statistics. Are a rising percentage of users accessing the website through mobile devices? Is there an important type of patron that you want to reach through a mobilized website, such as a large youth population? Or you may simply want to launch the first step towards a mobile website, knowing that it's an important direction to be moving toward even if your statistics are not that high yet.

Another consideration is the status of your library website. If you are in the process of revamping an outdated website, this can be a perfect time to introduce a mobile presence as well. If your website is fully established and usability tests show that users are happy and getting the information they need, your strategy will be quite different. In this case, you will most likely present established information and resources in ways similar to your existing website, rather than designing something brand new. Whether creating a new website or revamping an old one, you will want to ensure that your mobile website matches your overall website theme and library branding.

Just as in creating a full website or redesigning an old site, you need to be very clear in your mobile web strategy statement about the purpose of your website. We have already talked about knowing your website's audience, but just as important is knowing your users' top information needs when accessing your website as well as their browsing habits. Analytics can also be very helpful in this process. For example, Google Analytics saves content statistics, making it easy to identify the top resources accessed through your website. Not all information is appropriate or necessary to include on a mobile website, but you should be sure to provide for most of the top information needs of your patrons.

Analytics can also tell you the types of devices that patrons are using to access your site, but be careful to not rely too much on this information, as the mobile device market is very fluid and purchasing trends can change very fast. It is best to plan for a design that works well on many devices to avoid having to frequently redesign your mobile web presence because of changing trends.

Another consideration in strategizing a mobile web presence is your existing web content and how it fits onto the smaller screen of mobile devices. Simple designs are best when working with mobile design. Graphics should load fast, and you need to avoid large graphics and slideshows that may work well on full websites but not on small screens. You may also need to simplify your logo for your mobile presence. Avoiding multiple columns and extensive vertical scrolling is also very important. There are many elegant ways of designing mobile sites with expandable navigation. The key to a successful design is to decide on the categories you will use to direct patrons to the information quickly and clearly.

It is also important to consider that mobile devices will not have the flexibility that the use of a mouse affords. Touch technology limits the actions of users as they navigate your website, and this must be considered when designing buttons and other navigation features. Considerations such as these will ensure that your users have a streamlined experience and won't get frustrated when using your mobile website. Best practices for designing websites will be discussed further in chapter 7.

DESIGNING A MOBILE-FRIENDLY WEB PRESENCE

There are many considerations when beginning or expanding a mobile website. We will explore several of the common formats for delivering information to mobile users from your website in the rest of this chapter. Which format you choose will depend on your mobile strategy and individual library needs. Here are some of the formats to consider:

- Create clean website layout that translates well to smaller screen sizes
- Use a content management system that has a mobile template
- Create customized style sheets for different screen sizes or devices
- Create a mobile-only website
- Consider mobile apps
- Use responsive design for websites

When starting out, it may be enough to just keep your library website design clean and simple with a layout that will look good on a smaller screen. You will also need to keep graphics to a minimum and make sure your navigation works well with touch technology. Some guidelines for making regular websites display better on mobile devices include: not using tables or frames, using a single column layout, optimizing graphics, minimizing scrolling by using navigation tools, and keeping text to a minimum.

If a large portion of your users are mobile users, you may want to optimize your entire website for mobile use. Many resource vendors have gone this route with a one-column design, their search box taking up most of the screen, and minimal text on the home page. This is called a mobilized website, but most library sites are more complicated and require a more complex design.

Another way to get started in reaching your mobile audience is to use a content management system that is optimized for mobile devices, such as LibGuides or WordPress. LibGuides, the templating program from Springshare used by many libraries for publishing tutorials and library guides on the web, has "mobile-friendly" web pages that automatically adjust for optimal display on smaller screens and a mobilized menu system that appears in place of the home page on phone-sized screens. Embedded video in LibGuides does not use Flash, so it works fine on most mobile devices. There is also a module that you can purchase from Springshare that allows you to build mobile websites interactively without any programming knowledge. You can use the module to build a mobile version of your website interactively, and then the module produces the JavaScript code to add to your website so that mobile visitors will be automatically redirected to the new mobile-friendly version of your site.

WordPress, originally used for creating blogs, is becoming popular as a content management system for building complete websites, due in part to its clean layout and many available themes and plugins. Several WordPress themes and plugins can be used to make web content more readable from mobile devices. The simplest to use is WPTouch, which strips out the design elements from a website and leaves the content in a format that is easy to read on a small screen. The WordPress Mobile Pack can be used to display blog posts in a mobilized format. Many WordPress themes, including the popular Twenty Eleven theme, have responsive design built in. There are now also several themes just for mobile sites. WordPress for iOS (figure 3.2) makes it easy to administer your blog from a mobile device.

For a slightly more complex mobile solution, but one that gives you more control over what your website looks like on smaller screens, you can create customized style sheets with custom cascading style sheets (CSS) for each screen size. You decide what size of screens to design for and set up separate style sheets for each size that will remove graphics and extra columns as well as other features problematic for mobile websites. The style sheets are defined and added in the header section of your website so that it can recognize the screen size of the device accessing your site and switch to the appropriate style sheet.

Mobile-only sites have been popular for a while. These m. or mobile sites are created with a different URL from your main website. You can also use scripts on your website that detect screen size or type of device and switch to the mobile site automatically. These separate sites are typically scaled-down versions of the full website with content appropriate for mobile users. You should be sure to include a link to the full site if your website automatically redirects to a mobile site since some users may want to see your full site and need to access services not included on the mobile site in some instances. You don't want to limit them from accessing your full content if needed.

If your user base is ready for it, consider creating a mobile app. Creating a full mobile app is getting easier. There are many Internet web tools available that specialize in making it easier to create simple to complex mobile apps. For simple designs, no coding experience may be necessary. If you have a lot

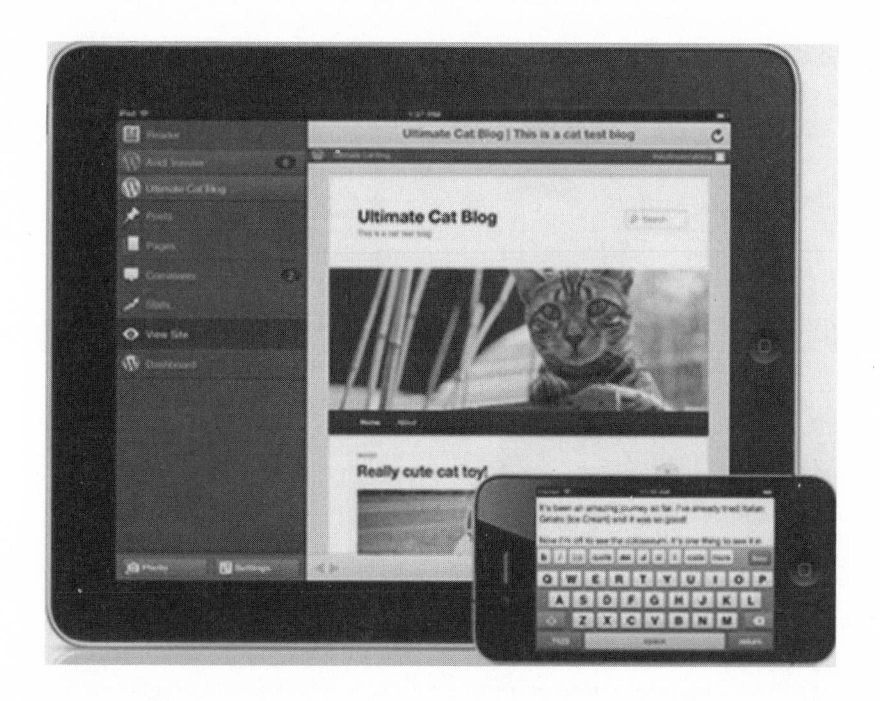

Figure 3.2. WordPress for iOS on the iPad and iPhone. *wordpress.com*

of iOS or Android users, consider creating a fully mobile app that users can download right onto their device. Users will not have to use the browser, and your library site will be easily accessible from the home screen. One of the drawbacks to this approach is having to create separate apps for different device operating systems and having to completely rebuild and re-release the app whenever you want to make a change. Some free mobile app creation tools include Winksite.com, Zinadoo.com, and MobilePress from WordPress. Another inexpensive option is MoFuse.com, which allows you to integrate GPS into your apps. This could be useful for a library location application.

There are currently two main types of mobile app layouts: the "grid" where icons are displayed across the screen like the home screen on an iPhone, and the "list" where menu items are stacked vertically down the screen. Figure 3.3 shows these two types of layouts. Both work well for touch navigation.

The responsive website design method is becoming increasingly popular due to the drawbacks of creating mobile apps. Using style sheets that detect different maximum or minimum screen sizes, you can create a website that automatically adjusts to smaller screen sizes. The navigation menus will

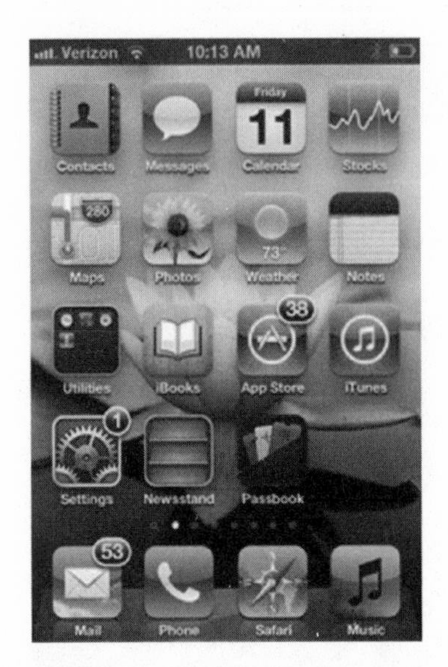

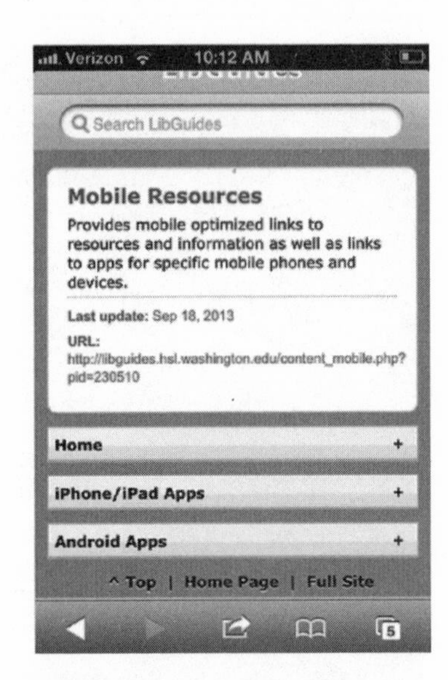

Figure 3.3. Grid layout on the iPhone and list layout used by LibGuides

change and graphics rearrange or even disappear as the screen gets smaller. You can customize the layout for as many screen sizes as you want with custom CSS. Responsive design divides the screen up into sections that take up certain percentages of the screen. These percentages change as the screen becomes smaller, and the sections can even be made to move to different areas or change in shape, size, and display with some coding or plugins such as Mobble for WordPress.

The newest type of design is called mobile first, and involves creating the mobile website first and then creating the desktop version. If you are designing a new site and most of your users are accessing your site from mobile devices, this is the way to go. In the future as users become more and more mobile, this way of designing websites will probably become more common.

After deciding what level of mobile access to provide, you will be ready for the actual design process. Remember that a mobile website is not just a scaled-down version of your full site. Provide only what content is appropriate for mobile users, and keep in mind the limitations of the small screen.

When working on the design of your mobile website, always start with analytics. What percentage of users are coming to your website from mobile devices? What level of design does this number warrant? What devices are they using? This information will also be important for testing your design later. If you have a large mobile audience, you may want to hire a specialized mobile web designer or app creator, but that is not necessary to get started as you can use any of the simpler methods of providing mobilized content described above.

TIP: Start small. Grow your mobile user base and then work up to more complex mobile solutions as the mobile access to your site increases.

CONTENT FOR THE MOBILE WEBSITE

Depending on what format your mobile presence takes, you may need to limit the content provided from your mobile site. This is especially true when creating a mobile-only site or an app. Using analytics, top content accessed from your website can be determined. Some content also works better on mobile devices than others. Text-heavy content usually does not work well, but quick facts and access to services and top resources are all important to include.

Top content to include on mobile library websites:

- Catalog and e-journal search boxes
- Links to popular resources: only those having mobile-optimized websites should be included
- Library hours
- Maps and directions or virtual tours
- Links to popular e-books, or e-book selection and download tools
- Chat and text with a librarian services
- Top resources: links to your most popular information resources
- Library news: customizable to user interests and including RSS feeds
- Study/conference room reservations: the mobile website is a perfect place to include links to your online room reservation system if you have one
- Podcasts/tutorials: consider creating podcasts or simple online videos of library use tutorials or recordings of library programs
- Customization features: allowing users to set their library account or choose their top resources is an advanced feature that appeals to mobile users
- GPS applications: since mobile devices have GPS built in, consider including a feature such as interactive maps, something that you can't provide on a desktop computer
- QR codes: these are useful for bringing mobile users to your mobile website

You can also integrate a mobile catalog, an e-journal reader, and other specialized applications into your design. Several vendors specialize in mobile design for libraries, such as Boopsie, which offers a fully integrated library system (ILS), library locator, and "ask a librarian" services, and can even integrate social networking with their product. Innovative Interfaces offers a mobile online public access catalog (OPAC) called AirPac, and LibraryAnywhere from Library Thing offers a mobile interface for your catalog and provides both iOS and Android apps. If you want to offer custom mobile access to your e-journals, Browzine offers a mobile app that displays your library's e-journal subscriptions in a readable format. You can download journal articles and save them in your mobile device's library for later reading. Browzine now has both Android and iPhone apps available. You would need to link to the Browzine app download from your mobile site and the app would be used separately on the mobile device.

DESIGN TIPS: Keep the interface simple and clean, minimizing the use of images. Avoid making users scroll. Keep content limited to specific features and resources that people want to access from mobile devices. Keep text to a minimum. Make sure your design works on multiple devices. Don't use Flash video; it won't work on some devices.

TESTING YOUR MOBILE DESIGN

After creating your design, it is crucial to test with multiple devices or at least with the ones most commonly used to access your website according to your web statistics. There are many tools to make this process easier. One tool to get you started is provided by MobiReady.com. This interactive site allows you to type in your website homepage URL and gives you a score on readiness for mobile and offers advice on how to improve your website for mobile devices. W3C MobileOK Checker offers a similar service at http://validator.w3.org/mobile/.

Making sure your newly designed mobile website looks good on multiple devices can be a challenge. Emulator programs can help with this task. One of the most popular vendors of mobile phones and tablets is Samsung. The company has developed an online testing center at Lab.dev that allows you to choose from the many Samsung devices and display an emulated screen where you can test the display of your site on a specific model. Symbian, Windows Mobile, and Java devices are available to choose from. For Android devices, the Android developer toolkit includes a virtual emulator that installs on your computer desktop. The iPhone Tester at http://iphonetester.com/ provides a simple online interface for testing your design on the iPhone.

Mobile Phone Emulator at www.mobilephoneemulator.com lets you choose from a wide variety of mobile phone types, including BlackBerry.

The last task in the design process is usability testing, which is extremely important. A poorly designed and inadequately tested site will turn away users who may never return to your site and go elsewhere for information. Mobile designers have recently developed heuristics for mobile library websites. One of these is a checklist developed by Tiffini Travis and Aaron Tay.[4] The checklist includes interface design questions, user characteristics, and content/purpose. Completing a heuristic checklist for your mobile design is crucial to its usability.

Paper tests can also be completed during the design process to help you decide between options. Storyboarding is a good way to mock up your mobile website design on paper to test before creating an actual website. With the release of your beta site, usability testing with actual users trying to complete tasks on your new site will help identify navigation and other issues that may need to be addressed before the final release of your mobile site.

Once you have created a mobile web presence, don't forget to advertise your new service. Create flyers and posters advertising your mobile service and distribute where your users will see them. Advertise your mobilized website in your newsletters and at library events and meetings. Consider posting QR codes that mobile users can scan to link directly to your mobile site or m. page. Get the word out so that your new mobile site will be discovered and used.

With the rapid growth of mobile usage, libraries cannot afford to wait any longer to consider creating a mobile presence for their website. No matter what your budget or expertise, there's a solution that's right for your library and your patrons. Get started by creating your mobile web strategy and exploring options for reaching out to your mobile users through your website. In the next chapter, we will discuss the next step after mobilizing your website—creating and using mobile apps in the library setting.

NOTES

1. Daly, Jimmy. "13 Impressive Statistics about Mobile Device Use." *EdTech*, March 13, 2013. http://www.edtechmagazine.com/higher/article/2013/03/13-impressive-statistics-about-mobile-device-use (accessed February 26, 2014).
2. Pew Internet. "Mobile Connections to Libraries." Pew Internet Libraries RSS, December 31, 2012. http://libraries.pewinternet.org/2012/12/31/mobile-connections-to-libraries/ (accessed February 26, 2014).
3. Carlucci, Thomas. "The State of Mobile in Libraries 2012." The Digital Shift, February 7, 2012. http://www.thedigitalshift.com/2012/02/mobile/the-state-of-mobile-in-libraries-2012 (accessed February 26, 2014).
4. Travis, Tiffini, and Aaron Tay. "Designing Low-Cost Mobile Websites for Libraries." *Bulletin of the American Society for Information Science and Technology* 38, no. 1 (2011): 24–29.

FURTHER READING

Aldrich, Alan. "Universities and Libraries Move to the Mobile Web." Educause Quarterly, June 2, 2014. http://www.educause.edu/ero/article/universities-and-libraries-move-mobile-web (accessed February 26, 2014).
Kim, Bohyun. "Improve Your Library's Mobile Website: Up from Start." ALA TechSource, September 9, 2013. http://www.alatechsource.org/blog/2013/09/improve-your-librarys-mobile-website-steps-up-from-start.html (accessed February 26, 2014).
McCollin, Rachel. "Creating Mobile-Optimized Websites Using WordPress." Smashing Magazine, May 24, 2012. http://mobile.smashingmagazine.com/2012/05/24/creating-mobile-optimized-websites-using-wordpress/ (accessed February 26, 2014).
Olsen, Dave. "Results from Higher Ed Mobile Website Technical Survey." dmolsen.com, February 1, 2011. http://dmolsen.com/2011/02/01/results-from-higher-ed-mobile-website-tech-survey/ (accessed February 26, 2014).
Warner, Janine, and David LaFontaine. *Mobile Web Design for Dummies*. Hoboken, NJ: Wiley, 2010.

RESOURCES

Analytics

Google Analytics: http://www.google.com/analytics/
Google Analytics for Mobile: http://www.google.com/analytics/mobile/
Open Web Analytics: http://www.openwebanalytics.com/
StatCounter: http://statcounter.com/
WebSTAT: http://www.webstat.com/
WebTrends: http://webtrends.com/

Mobile Tools and Examples

AirPAC from Innovative: http://www.iii.com/products/airpac.shtml
Boopsie: http://www.boopsie.com/
Browzine: http://thirdiron.com/browzine/
LibGuides from Springshare: http://springshare.com/libguides/
Library Anywhere from Library Thing: http://www.librarything.com/blogs/thingology/2010/01/library-anywhere-a-mobile-catalog-for-everyone/
M-Libraries Wiki: http://www.libsuccess.org/index.php?title=M-Libraries
Mobble: http://wordpress.org/plugins/mobble/
MobilePress: http://wordpress.org/plugins/mobilepress/
MoFuse: http://mofuse.com/
Winksite: http://winksite.com/site/
WordPress: http://wordpress.com/
WPTouch: http://wordpress.org/plugins/wptouch/
Zinadoo: http://www.zinadoo.com

Testing

Android developer emulator: http://developer.android.com/tools/help/emulator.html
iPhone Tester: http://iphonetester.com/
Mobile Phone Emulator: www.mobilephoneemulator.com
MobiReady: http://ready.mobi/launch.jsp?locale=en_EN
Samsung Remote Test Lab: http://developer.samsung.com/remotetestlab/rtlDeviceList.action
WC3 MobileOK Checker: http://validator.w3.org/mobile/

Chapter Four

All about Mobile Apps

It seems like there is an app for everything now. Both the Apple Store and Google Play achieved over 1 million apps available for download during 2013. While smartphones have full access to the Internet and can bookmark sites and even save links to the desktop computer, users want shortcuts to information and interfaces to their favorite social networking sites available at the touch of a finger. Apps provide quick access to favorite web applications, customized for a small screen and fast loading on mobile devices. Apps can store login information to make it quick and easy to look up information, check email, connect to friends, or connect to shared documents saved in the cloud. Although many websites are becoming "mobile optimized" for easy browsing from small screens, apps are preferred over the full website in many cases.

A 2013 poll by the Nielsen Company concluded that while smartphone use is increasing, so also is the use of apps for the most popular services (figure 4.1).[1] Users of top services such as Google, YouTube, and Facebook have seen decreases in web access while simultaneously seeing large increases in their app usage. Nielsen reports that the top rated Facebook app saw a 27 percent increase in app usage and a 16 percent decrease in web logins. Nielsen estimates that on average, 103 million users access the Facebook app each month. Other top-rated apps are Google Search and Google Play at over 75 and 73 million uses on average each month. YouTube comes in fourth at almost 72 million uses a month, and Google Maps is fifth with 68.5 million average uses. Gmail, Instagram, Apple Maps and Stocks, and Twitter are the other top apps, according to Nielsen.

WHAT ARE APPS?

Apps are custom interfaces to information or scaled-down access to a selection of content from a full website. Although any website can be saved as a browser favorite or added to your smartphone home screen, apps provide mobile-optimized access to information at the touch of a finger. Some apps require Internet access to retrieve the full content, while others will run even without access to the Internet. Apps are customized to load faster and run better on mobile devices than on a website. Scrolling is minimized and access is usually limited to the most popular features.

In a 2010 survey, Nielsen reported that the most popular category of app downloads was games, followed by music and social networking apps.[2] News and weather apps were also very popular, along with video apps, food and entertainment, sports, finance, shopping, and productivity. Some popular apps such as the Facebook app are actually preferred over the full website version. Creative touch menu options that allow menus to open and close at the touch of a finger make it possible to access the full content available without too much scrolling.

When installing apps on your mobile device, you must log in to your personal Apple or Google account to download the software, even if it is a free app. Some apps such as the Kindle app allow you to download and use the app on multiple devices, but many apps can only be downloaded and installed once. Because mobile devices are personalized, it is hard to share apps in a business environment. Apple has created the Volume Purchasing Program to allow businesses to purchase multiple copies of apps and install

Rank	App	Average Monthly Users	Change from 2012
1	Facebook	103,420,000	27%
2	Google Search	75,984,000	37%
3	Google Play	73,984,000	28%
4	YouTube	71,962,000	27%
5	Google Maps	68,580,000	14%
6	Gmail	64,408,000	29%
7	Instagram	31,992,000	66%
8	Maps (Apple)	31,891,000	64%
9	Stocks	30,781,000	32%
10	Twitter	30,760,000	36%

Figure 4.1. Top smartphone apps of 2013. *Nielsen*

them under multiple individual accounts. Google does not yet have this option available from the Google Play store.

When deploying tablets or phones to employees or for a mobile device checkout program in libraries, it's difficult to standardize and setup multiple devices as is typically done for PCs or laptops because of the reliance on personal accounts for apps. The Apple Volume Purchasing Program makes setting up mobile devices for employees in a business environment a little easier. Problems arise when libraries want to checkout tablet computers to patrons who want to log in to their personal accounts on cloud app or social networking sites. One way to work around the individual account issue is to create a master iPad loaded with all the apps needed for your library program on one master iTunes account. Then, you can back up the master iPad to iCloud; this creates a copy that can then be used to set up each iPad in your checkout program. Apple allows you to choose to set up a new iPad from an iCloud or iTunes account rather than create a new account for each device. After setting up the new iPad, you must rename it from the settings menu so that it has a unique name and then set it to back up to iCloud so it can be restored after each use.

Apps loaded on an iPad for a library checkout program could include apps for productivity, resources, and social networking. Apps for information resources will usually allow you to log in with a generic account on multiple devices. Some may be more restrictive, especially more expensive apps. Productivity and social networking accounts such as Google Drive or Facebook require a personal login to access your files and personal account. This can be a problem on shared devices, because apps will many times store the user login information when someone logs in with their personal account. With iOS, some apps can be cleared from the settings menu with a reset option, but many cannot. To work around the limitation of saved accounts on iPads, the device can be reset from the settings menu, and then the original iCloud backup can be loaded back on the iPad with fresh accounts without login names and passwords before the device is loaned out again. Android devices can be backed up to the Google Cloud for a similar setup.

Another consideration when loaning out mobile devices in libraries is cleanliness. Especially in a medical or educational environment, the spread of germs with shared devices may be of concern. One way of dealing with this issue is to use a cleanable case and screen cover on all devices. Also, devices that are stored for a while between uses should be powered off to avoid running down the battery between uses. Even when powered down, if a tablet is stored for a long time, the battery will need to be recharged. Mobile charging station tables or kiosks are starting to appear in some libraries for mobile device users who need to recharge their batteries. Public charging stations are also appearing on city streets and other public places, but beware that hackers are able to attach smaller computers to these stations, which can download a virus onto your mobile phone.

CREATING APPS FOR THE LIBRARY

In a 2010 American Library Association policy brief entitled "There's an App for That!" the author outlines several compelling reasons why libraries should consider providing customized mobile access to their services.[3] Apps provide access to library services anywhere, anytime. Use of library resources such as collections of e-books, e-journals, and videos can be promoted and expanded by providing mobile access. Some other important services for which patrons frequently request mobile access include the library catalog, access to their library account, placing books on hold, and library reference services. Concerns that were raised in the report and need to be addressed when planning a mobile app included security and privacy, problems with the use of location awareness, and integration with social networks, as well as issues with digital content licensing.

In a 2012 survey done by Birmingham City University on the implementation of mobile technology in libraries, only 19 percent of respondents had implemented a mobile app for their library.[4] Barriers to implementing mobile technology included resource constraints, infrastructure limitations, lack of technical support, and lack of knowledge about mobile technologies. Suggestions for overcoming these barriers included finding lower-cost solutions, hiring knowledgeable staff into new positions, reassigning and training staff members, and collaborative partnerships with knowledgeable businesses and IT organizations. Although initially very expensive, creation of apps is becoming cheaper and more accessible to libraries on a budget due to the introduction of app creation software and vendor-created apps.

The opportunity to reach out to an ever-growing new audience of users of mobile apps should be a compelling reason for libraries to consider developing their own apps. An app can be created to access the most popular library resources and services. An app can also allow access to electronic online materials such as e-books and e-journals. Libraries might also consider creating apps to highlight important community projects or outreach initiatives. Any project that might result in a website can be considered for a possible app project as well. There is now even software available that will convert a simple website to an app format. For example, Uppsite is a plugin that will convert a WordPress website to an Apple, Android, or Windows phone app in a couple of minutes.

Before creating an app you will want to consider the purpose your app will fulfill, the users you want to reach, and the resources and services you want to include, as well as any graphical elements to be included such as logos and color schemes. Another way to plan for development of an app project is to think of what problem the app solves or solution it will provide to your users. For example, you may want to create an app that allows your users to download and read library e-books through their library account from anywhere they have Internet access. You will need to design the interface or "views" that will allow the user to interact with your app in order to accomplish this task. The user interface should be simple, efficient, and avoid confusing the user. Storyboarding, which is a series of illustrations displayed in sequence, can be used to visualize your app project, detailing the steps the user must take to complete a task. The storyboard can then be used to create views within the app. Storyboarding can also be used with test users in usability testing your app concept before creating the actual app.

Apple Development

Anyone with some programming background can develop apps. Apple has a basic tutorial at http://developer.apple.com along with a lot of other helpful resources for developing iOS apps. In order to develop an iOS app, you will need a Mac computer running OS 10.7 or later, Xcode software, and iOS SDK. Xcode is Apple's free development environment software or integrated development environment (IDE). It includes a code editor and a user interface editor, tools that make iOS app development easier. The iOS SDK is the Apple software development kit, which contains ready-made tools and frameworks for creating features used in apps, such as menus. The iOS SDK is included with Xcode.

Android Development

The Android developer site also has a tutorial for building Android apps at http://developer.android.com/training/basics/firstapp/index.html. Android development requires the Android SDK and accompanying development tools. Another option is to use the Eclipse development environment with the Android development tools plugin. Just like developing apps for Apple devices, you will create views that will walk users through tasks to be completed using the app. The development tools help create the programming code needed to create the app views and make them work with the Android OS and touch screen technology.

When creating an Android app, you need to consider which versions of Android you will support. Like any open-source software, development of the Android OS is constant and new versions come out often. It is a good idea to make your app backward compatible so that mobile devices running earlier versions of the Android OS will be able to run your app. Since 2009, Android versions have been named in alphabetical order with a dessert-based theme. Early versions were code-named Cupcake and Eclair. The latest versions are Ice Cream Sandwich, Jelly Bean, and most recently KitKat, released in November 2013. No one yet knows what the next version will be called but it's probably safe to say it will start with the letter L and follow the dessert theme.

Because the Android system runs on so many different hardware products, unlike iOS which only runs on Apple products, there are many versions of the Android operating system currently in use at a time (figure 4.2). According to the Android developer website, there are currently six main versions running on Android phones: Froyo, Gingerbread, Honeycomb, Ice Cream Sandwich, Jelly Bean, and KitKat.[5] It is important to know which version you are developing for when creating an Android app, and the Eclipse development environment with Android plugin lets you choose the minimum and maximum versions you will support.

When developing apps, no matter which operating system you use, it is important to test your app often during the development process. If you are developing for Apple, you should test on the iPhone, iPad, and also the iPod Touch. Xcode provides support for building universal apps that will run on all versions of Apple products and will adjust to the different screen sizes. You also have the option to build an app for a specific device such as just the iPad, if you choose. When developing an Android, Windows Phone, or BlackBerry app you can choose to test it on an emulator program. There are many of these free programs to download on the Internet and these are also available from phone manufacturer websites. An emulator runs in a browser

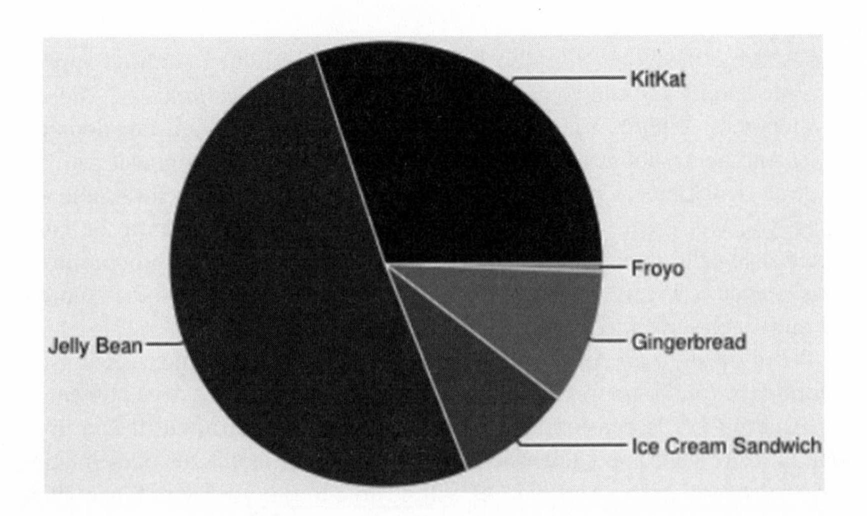

Figure 4.2. Android versions currently in use. ***Android developer website***

window and simulates the experience of using an app on the emulated device.

It is also possible to create apps that run across multiple platforms. Be aware though that the app may not be using the full functionality of the device unless programmed specifically for that operating system. The development environments for each device mentioned above use specific programming languages native to the device. Apple iOS currently uses Objective C language with Cocoa, Android uses Java, and the Windows Phone uses C# with .Net programming language. HTML5 can be run on all three platforms as can code written in the C++ programming language. Unfortunately, it is much harder to create apps using just programming language without development tools. Using a development environment with plugins to create your code for you is much easier, and there is no need for advanced training in programming languages to create a simple app.

Some best practices for creating apps that are easy to use and popular with users include:

- Minimize scrolling by dividing content into pages or views
- Use collapsible menus to expose or hide information from the screen
- Place the most popular and most highly accessed resources at the top of the screen where they can be easily seen and used
- Brand your app with your library log and design template for recognition and marketing

Planning ahead during the design process and pre-testing with storyboards can make a big difference to the final product. There are also services and companies available to help build custom library apps or services. These services can create complex and well-designed apps for libraries that integrate with your catalog and other services, but they can be expensive. Services to help you build apps for your library will be discussed further in chapter 6.

RECOMMENDING APPS TO LIBRARY USERS

Many libraries keep recommended lists of apps for their users. Many library websites include mobile resources pages, which may include instructions for reading library e-books from various e-reader apps; lists of information resources with apps available to download; or guides to apps on a specific subject or topic. Depending on the type of library and user needs, most libraries should at least consider listing recommended apps from their websites. Knowing that a portion of your users will access your online resources from a mobile device, it will be important to include instructions for downloading apps available for your most-used resources and databases. Many database vendors and specialty resources are releasing apps to reach the mobile audience. Many of these require a personal account to download, so that the vendor can get contact information from users. Most of these apps are available free of charge.

Subject librarians may want to develop lists of apps for specific resources in their subject areas. There are some excellent free or inexpensive apps available for many different subject areas such as medicine, astronomy, physics, music, foreign languages, and history. The Apple Store and Google Play both list apps by topic to make it easy to search for new apps. There are also many articles in library and subject-area journals that list recommended apps for subject areas. Websites that are good for searching for apps include Skyscape for medical apps on any platform, AndroLib for Android apps, and Appolicious for both iOS and Android apps. There are even apps for recommending apps. Librarian-selected apps for resources and e-book reading can also be loaded onto iPads for checkout and use in the library.

Keeping up with new apps and developments in mobile devices can be difficult. Joining email lists for mobile topics and perusing websites devoted to app reviews can be helpful. Many library conferences now include program tracks on mobile technology where you can learn from other librarians and professionals in the field. Conference exhibits are also a great place to catch up with the newest developments from information resources vendors, many with mobile apps available or in development. In the next chapter we will explore the types of information seeking done with mobile technology, and in chapter 6 we will explore specific apps designed for use in libraries, including resources to create apps as well as apps for reading e-content plus productivity apps and other useful resources.

NOTES

1. "Tops of 2013: Digital." Nielsen.com, December 16, 2013. http://www.nielsen.com/us/en/newswire/2013/tops-of-2013-digital.html (accessed February 27, 2014).

2. "The State of Mobile Apps." Nielsen.com, June 1, 2010. http://www.nielsen.com/us/en/newswire/2010/the-state-of-mobile-apps.html (accessed February 27, 2014).

3. Vollmer, Timothy. "There's an App for That! Libraries and Mobile Technology: An Introduction to Public Policy Considerations." American Library Association OITP Policy Briefs, no. 3, June 2010. http://www.ala.org/offices/sites/ala.org.offices/files/content/oitp/publications/policybriefs/mobiledevices.pdf (accessed February 27, 2014).

4. Dalton, Pete, Jo Alcock, Yvonne Graves, and Sukhvinder Kaur. "Report on Current m-Library Activity." Mobile Technologies in Libraries RSS, February 13, 2012. http://mlibraries.jiscinvolve.org/wp/2012/02/13/report-on-current-m-library-activity/ (accessed February 27, 2014).

5. "Dashboards." Android Developers, n.d. https://developer.android.com/about/dashboards/index.html (accessed February 28, 2014).

FURTHER READING

Zickuhr, Kathryn, Lee Rainie, and Kristen Purcell. "Library Services in the Digital Age." Pew Internet Libraries RSS, January 22, 2013. http://libraries.pewinternet.org/2013/01/22/library-services/ (accessed February 27, 2014).

RESOURCES

120,000 Apps in BlackBerry World (Spoiler: 47,000 Made by One Developer): http://allthingsd.com/20130821/120000-apps-in-blackberry-world-spoiler-47000-made-by-one-developer/
Android Developer: http://developer.android.com
Androlib: http://www.androlib.com/
Apple Developer: http://developer.apple.com
Apple Volume Purchasing Program: https://developer.apple.com/support/appstore/volume-purchase-program/
Appolicious: http://www.appolicious.com
Skyscape: http://www.skyscape.com
Storyboarding Tips: Step by Step: http://webtoolsuw.wordpress.com/2010/05/18/storyboarding-tips-step-by-step/
Uppsite: https://www.uppsite.com/

Chapter Five

Information Seeking with Mobile Devices

The physical library is no longer the default place where people go when they need information. In a study on where people go for information, published in Educause Review, the authors found that "interviewees mentioned Internet resources, such as search engines and social media sites, far more often than physical places."[1] Survey respondents remarked that they had not been to a library in years, since they do much of their journal reading online. Many survey respondents did not even associate the online resources they were using with an actual library. The study authors recommended that libraries create easy-to-use online interfaces or portals to information, and entice people to use them with marketing and promotion. Branding library portals will better associate online resources with the library that provides the resources.

The authors also recommended that the library portal provide a variety of tools for finding information instead of locking users into one method, to allow for a variety of searching styles. The portal interface design must be simple and convenient to users because people will abandon complex library interfaces for easier search tools such as Google. Barriers between finding and accessing full text materials must also be removed. For example, library interfaces must clearly indicate if full text access is available early in the search process. Finally, the authors also recommend promoting and marketing library search portals so that users know they exist. Librarians know that library resources provide much better access to quality information than Google, but users who are looking for quick answers need to be convinced. The library portal should be mobile-friendly, or a mobile app should be created to entice mobile users away from Google.

Changes in the way people access information are bringing correspondingly big changes to the library field. Mobile devices allow users to take the library with them. This reduces the need for physical space for books and opens up more space for studying, meetings, and other activities in the physical library space. Libraries have been dealing with electronic resources, such as e-journals, e-books, databases, and websites, for many years now. The mobile revolution means libraries must now also deal with mobile websites and mobile devices and apps. It becomes a major challenge to libraries when researchers, students, and other information seekers purchase their own electronic subscriptions or apps because the library does not offer mobile access to resources users want. Libraries and librarians must adapt to the new mobile world and begin offering mobile-ready resources and portals to information, or risk becoming obsolete.

Many sources reporting on mobile trends remark on the growing use of mobile apps over mobile web browsing (figure 5.1). Mashable reported that mobile app usage increased by 115 percent in 2013.[2] TechCrunch reported that app usage dominates mobile web usage. In 2014, 86 percent of mobile device usage time was spent on mobile apps while only 14 percent of the time was spent using a mobile web browser.[3] In a May 2014 article in the *Independent*, Amazon was quoted as reporting that for every 100 physical books purchased, 114 e-books were downloaded.[4] This number would be even higher except that the number of downloads of free books was not included in the report. While print books still outsell e-books worldwide, the gap is closing. Amazon also reports that while e-books are cheaper than physical books, Kindle users are buying on average four times as many books as they did prior to owning a Kindle. The report indicated that the e-book format encourages people to read more and the statistics show that the format is convenient and conducive to multiple reading styles.

MOBILE INFORMATION SEEKING IN OUR DAILY LIVES

A 2010 American Library Association Library Brief points out that mobile technology is increasingly being used in most aspects of our daily lives.[5] Mobile technology provides access to medical information, facilitates access to information during emergency and disaster situations, and offers breaking news and updated financial information. People are able to conveniently access weather reports, check bus schedules, look up movie times, and check their bank accounts from mobile devices, whenever and wherever they need access to that information. Apps also keep us entertained and connected to friends and family, and help us navigate our world. Similarly, the search for information can be integrated into daily work and personal life. Easy-to-use library search portals should be available for checking a user's library ac-

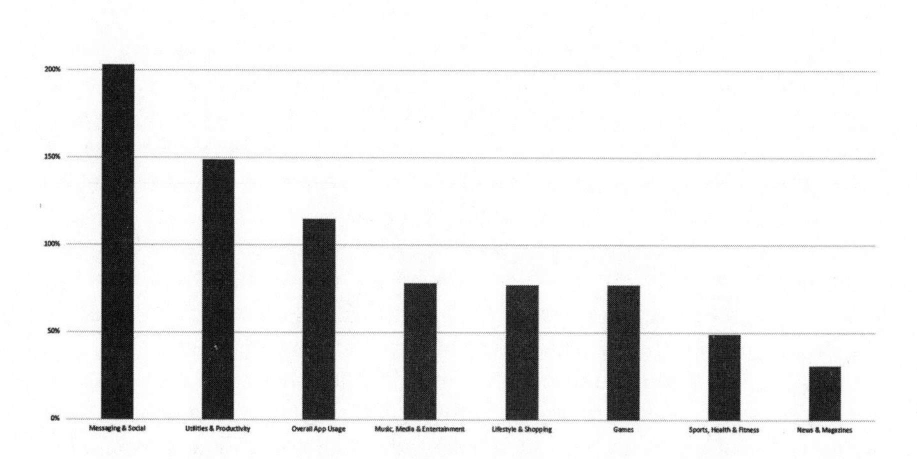

Figure 5.1. Growth of app usage by category in 2013. *Flurry Analytics*

count, finding library hours, looking up a book or reference, and downloading a book in e-format.

The types of information that people search for using mobile devices and mobile apps can be divided into the following basic categories:

- Education: access to course information and schedules, as well as online coursework
- Recreation: movie times, concert information, and schedules for parks and recreational activities
- Work: access to online calendars, email, and productivity applications
- Travel, directions: GPS, maps, and directions
- Reference: quick access to information and answers to questions when they come up in our daily lives
- Health and wellness: apps for keeping track of exercise and promoting health and wellness are increasing in popularity
- Sports and news: most top news and sports agencies provide mobile access to their information
- Games: recreational apps top the list of most popular apps
- Social: social networking apps such as Facebook, Twitter, and LinkedIn are some of the most popular downloads
- Home: apps for finding homes for sale or rent and access to home security and remote control utilities are now available
- Shopping: many top retailers including Amazon provide mobile access to their products

- Financial: access to instant, up-to-date financial information from mobile devices is also in demand

Reporting on global mobile usage for 2012, MobiThinking.com stated that according to Google, mobile searches quadrupled during the year, making one in seven searches mobile.[6] Top searches via a mobile device worldwide included accessing financial information, news, sports, weather, search, retail, travel, and reference information. One in every three recorded searches was for looking up local information and businesses and activities. Twenty-five percent of all search queries were done from mobile devices. There are two ways a library user might look for library information from a mobile device. The most common way would be to use the browser and a search engine. If your website has a mobile-optimized website, the mobile user will be satisfied and find the information needed. If not, they will likely look for the information elsewhere in the future. The second way a mobile user searches for information is to download and install a mobile app for a commonly used source of information. This option has still not been adopted widely by businesses and service industries such as libraries, although it is the easiest way to get information to your users. It is getting much easier and cheaper to create a mobile app, and large libraries with web statistics that show many users accessing the website from mobile devices may want to consider this option.

MOBILE INFORMATION SEEKING IN THE WORKPLACE

Mobile apps are increasingly being used in the workplace. An AT&T small business survey found that the most popular types of apps used in business settings included GPS/navigation (75 percent of users), location-based services (43 percent), document management (35 percent), and social media marketing (32 percent).[7] Other uses were for tracking expenses, delivery, and time management as well as mobile marketing and advertising. The top reasons reported for using mobile apps for businesses were time saving and increased productivity. Today's workers as well as students, who are the workforce of the future, expect to be able to access information from mobile devices. According to the AT&T study, the top advantages reported for mobile apps over websites were time saving and productivity. Generating new business and the coolness factor were also reported as advantages of having a mobile app.

Bloomberg reports that the Apple iPad is becoming a standard business tool. In a survey, about half of managers using an iPad reported they always use the device at work and another 40 percent say they use it sometimes at work. Seventy-nine percent of the managers surveyed used the iPad for business when out of the office.[8] A 2013 InfoWorld article reported that now that the iPad can be made HIPAA compliant, it is becoming the tool of choice for accessing patient information and keeping doctors and nurses connected to treatment information at the bedside.[9] Since the iPad mini now has a 10-hour battery and fits in a pocket, many businesses are providing these tablets for their employees to stay connected, access business and patient records, share documents with clients, and do work in the field instead of remaining tethered to physical office location with a desktop computer.

It is easy to recognize from these and other studies on the rising use of mobile technology that many of our library users carry mobile devices with them everywhere, and now many are also starting to use them in business settings, not just for personal use. Librarians need to pay attention to these trends and plan for implementing mobile technologies to better serve our increasingly mobile clients. This can be a huge challenge when libraries are also facing decreasing budgets and constantly increasing electronic resource pricing. It is important to plan carefully and carry out needs assessment activities to identify the types of information typically accessed by users of your library, and the devices most commonly used, which can be determined by the use of web analytics products such as Google Analytics or WebTrends.

Mobile devices can be used to support librarians in their roles of providing information, teaching and learning, and accessing research materials. As more resources are provided online and fewer patrons actually step foot in the physical library, the need for librarians to be more mobile and reach out to our students, researchers, clinicians, and citizens is growing. Training in the use of mobile devices for librarians and staff is crucial for successfully implementing mobile initiatives in libraries.

INTRODUCING MOBILE DEVICES IN THE LIBRARY

In the fall of 2011, the University of Washington Health Sciences Library (UW HSL) launched a pilot project to explore the use of tablet computers in the library. We hoped to become more mobile by integrating tablet computers into our daily work. Initially, we purchased two iPads and one Samsung Galaxy Tab. During the pilot project, we addressed three specific areas of need: providing health information to community members by participating in community health fairs, providing clinical librarians with mobile computing tools needed while attending hospital rounds, resident reports, and medical consults, and supporting the UW School of Medicine in developing a long-term mobile device strategy for medical education. We found that tablets were perfect for one-on-one instruction or small-group demonstrations during community outreach sessions due to the small size, long battery life, and touch-screen technology. The tablets also permitted clinical librarians to access information resources at the point of need in hospital and clinical settings. It also allowed us to work more closely with the School of Medicine in evaluating the use of mobile devices and in evaluating apps for use in medical education.

The initial pilot project was very successful and we learned some important lessons. First, we learned that a mobile device was very useful for librarian outreach, particularly in a clinical setting. We also discovered that the tablets were difficult to share between librarians. They are highly customizable for individual use and setup using an individual personal account with Apple or Google. Although we created a generic account for our initial testing, each librarian who used the device had different apps to install according to his or her area of specialty, and each person organized apps and utilities differently. We also learned that mobile devices are constantly changing, and the devices we purchased in 2011 were obsolete in 2012 with the release of the iPad2.

Major lessons learned include:

- Tablet computers can be used as an essential tool for medical librarians in their daily work, although they cannot yet totally replace a desktop or laptop computer.
- Tablet devices are optimally used as personal devices rather than being shared due to the built-in personalization features.
- The e-book market is constantly changing and vendor platforms are widely varied, making it difficult to access some e-books from tablet devices.
- Mobile devices are constantly changing, making it difficult to keep up with the new technology.

Following the success of our initial pilot project, and with the realization that mobile devices are highly customizable and not easily shared, we decided to find funding to purchase tablet computers for each librarian who wanted to use one. We gave each librarian the option of choosing an Apple iPad or Samsung Galaxy Tab. Although the iPad was the most popular device, we wanted to make sure that we stayed familiar with the Android operating system as well. We were able to purchase seven more Apple iPads and one additional 7-inch Samsung Galaxy Tab for HSL librarians. This enabled all HSL librarians to have an individual tablet computer, which they could customize to fit their daily tasks and workflow needs. HSL computing staff members set up the tablets and worked on creating a "master" image to streamline setup and for rebuilding the tablets when needed. A list of useful productivity and clinical apps was compiled and installed on the tablets. Librarians attended training sessions on the setup and use of the tablets, and several follow-up workshops were given to demonstrate innovative uses and provide support. An existing mobile LibGuide for users was updated to reflect new clinical apps that were found to be useful by librarians participating in the project. Methods of downloading e-books and PDFs to tablets were explored, enabling librarians to better answer faculty and student questions about e-books. Computer lab teaching classrooms were equipped with VGA cables to make it easier to connect the tablets to large screens for teaching sessions.

A survey of all librarians participating in the project was conducted after the initial training period. Survey results indicated overall satisfaction with the devices. Some librarians found them invaluable, but others indicated they were still evaluating the usefulness of the devices. Top uses for the devices included reading email and accessing documents, browsing the web, taking notes, reading e-texts, answering reference questions, and testing new apps to access resources or provide information. All librarians agreed it was important to keep up with tablet technology to best serve our users. The project was very successful in introducing mobile technology to librarians, putting mobile technology directly in their hands so they could learn on the job. We are continuing to provide tablet computers to all librarians in our library and have recently purchased iPad minis to evaluate the usefulness of the smaller-size device. We are also evaluating the Microsoft Surface computer and keeping up with developments in Android devices.

PROVIDING MOBILE ACCESS TO INFORMATION

Beyond merely becoming familiar with and using mobile devices in our daily work, librarians should also be providing mobile access to services and information. The first thing that libraries usually think about providing mobile access to is the library catalog. Other ways that libraries can provide mobile access to information include sharing collections of mobile apps, providing access to e-books, using mobile chat and roving reference services, initiating mobile lending services, and providing mobile access to databases through vendor apps. Deciding which of these services should be provided in a particular library depends on users, services, and budget. Users may not be particularly interested in accessing the library catalog from a mobile device, but other services may be more popular. Assessment needs to be done to determine the needs of library users and in what contexts they will be accessing your library services from a mobile device. Services that are already popular in a library are a good starting point for investing time and money on making them more mobile-friendly. Librarians also need to be comfortable with the technology and receive training in order for any mobile program to be successful.

Most libraries provide some form of reference services. Mobile devices are being used in libraries as roving reference tools. At Southern Illinois University, librarians have experimented with using iPads for providing reference services in the library building.[10] Since the wireless signal is excellent throughout their library, librarians are able to move around the library answering questions at the point of need. They can also answer chat questions and email from anywhere in the building, freeing them from being tied to the reference desk. Working with library patrons throughout the library, librarians can look up information in the catalog, browse the Internet, access multimedia collections, and demonstrate how to use library resources without patrons having to leave their study spaces and find a librarian at the reference desk.

Another popular use of iPads in libraries is for equipment loan programs. At Briar Cliff University, iPads can be checked out for use in the library as well as outside the library.[11] The iPads are set up with generic accounts and preloaded with many useful productivity, social, and information-finding apps. After each use, the iPads are reset to clear out any account information so that the next user can access cloud apps with their own account information. Unfortunately, users cannot download new apps or e-books, but the iPads are loaded with prepurchased apps and free e-book collections. Fees are charged for late return or lost equipment, and there is a clear policy for usage that lists fines and replacement fees. Technology support staff members are essential to this type of program due to constant required maintenance and updating of the equipment.

Librarians are often asked to recommend apps for finding database information. Apps provided by resources vendors vary in their availability, quality, and functionality. While many resources have apps available, some rely on mobile-friendly or responsive website designs. Most vendors require users to create a free login account to access their mobile app. A few make downloading very difficult by requiring multiple accounts and logins before you can access the resource. Several require the initial account to be created from within the library's IP range to authenticate the user. In addition to apps for specific library resources, there are also many cloud and productivity apps to facilitate collaboration, access, and sharing of information on tablet devices. Providing librarians with iPads allows them to test the utility and quality of apps in order to recommend them to library users.

RECOMMENDING APPS TO LIBRARY USERS

Many database vendors now provide mobile access to their search interface. Some provide a mobile-friendly website, while others create iOS or Android apps or both. Only a few create apps for all platforms. Mobile-friendly websites are easier to maintain than apps, which may have to be updated and re-released through the Apple Store, Google Play, or other app provider. While a mobile-friendly website may work with most browsers and on all devices, separate apps need to be created for each mobile operating system. Some vendors require the creation of a login account for access to their apps. Others require authentication through an institutional membership. Many libraries provide lists of apps for library resources and other useful apps from their websites. This service is popular at our library and one of the more highly accessed pages on our LibGuides site.

When selecting mobile apps for recommending to users or creating lists of helpful apps on your library website, it is important to start with a selection policy. If the list of apps is long, you might want to divide it up into pages or topics in order to make it easier for people to find what they are interested in. In creating a selection policy, decide what kinds of apps will be most useful to your library patrons and what guidelines you will use to decide whether or not to include an app on the list. Some selection criteria and questions to think about when selecting apps are listed here:

- Platform: Is it available in iOS and Android? What about other devices?
- Connection: Is the information available without Internet access?
- Cost: Is it free? Is registration required, or an institutional account needed?
- Stable version: Is it hosted by a reputable company or institution? Do all the links work, and is it maintained and up to date?
- Easy to use: Is it easy to find information, and are navigation features clear?
- Image quality: If images are included, are they high resolution?
- Works well without bugs: Are there bugs in the software or does it restart without warning?
- Essential resource: Is this an app for one of the library's essential resources?
- Commercial-free: Are there ads displayed, and are they distracting? Is the app biased by commercial support?
- Scope, volume, and quality of information: Is the amount of material included substantial, and does it cover only one limited topic or multiple subjects? Is the information provided of the highest quality?
- Usefulness at point of care or for research: Is the app useful for clinical care diagnosis and treatment or support research methodology?

There are many quality free apps from government and institutional sources. The National Library of Medicine has many high-quality, free apps available for download. A listing of these apps can be found at http://www.nlm.nih.gov/mobile/. Be advised that free apps that become popular may be bought up by businesses and no longer be offered free of charge. There are many apps that have a free version with limited content, and a "pro" version with full content for a fee. A good way to find new apps to preview for library user lists is to review the iTunes or Google Play stores under the specific topic, such as medical apps. Check out app review and hosting websites such as iMedical apps and Skyscape for medical-related apps. Check out mobile guides from other institutions for suggestions as well. There are even apps for recommending apps, such as AppGrooves and Appreciate for the Android market.

APPS FROM YOUR LIBRARY

If your library website statistics show a high level of mobile access to your website, you might consider creating your own app for mobile access to your library information. Types of information to include on a library mobile app may include access to your library catalog with loan and renewal information; hours and directions to the library; contact information via instant messaging, chat, or phone; links to mobile-friendly databases; links to library social networking sites; maps of library locations; information on computer lab and study room availability and online reservations; library news and events; and instructions for downloading e-content. There are vendors such as Boopsie that help create apps for libraries. Boopsie builds apps that are branded with your library's logo and colors. They also cover apps for all the major platforms. Boopsie works with most major catalog systems to integrate your library catalog into the app with search, account access, renewal, and even checkout options. Library information such as hours, maps, events and news, and access to chat service can also be included in the app along with access to e-books. Once you have created your library app, be sure to promote it from your website and other venues so that users know about it and use it.

When creating a library app, it is important to consider not only the services you provide and want to promote, but also use case scenarios that apply to your specific users. Consider all your library services and decide which ones would benefit most from a mobile point of access. An OCLC Research article cautions potential app creators against having too-high expectations when creating an app.[12] Users do not need to access all of your services from a mobile device, so don't try to provide them with everything. Creating an app without considering the contexts in which it will be used and the corresponding services that user will want to access may lead to disappointing results. An example of an application that lends itself well to a mobile interface is the North Carolina State University Libraries' coffee bar webcam. Students can access the library mobile app and see how long the line is at the coffee bar located in the library before deciding whether or not to get coffee between classes. When assessing the app, developers found that access to this page of the app accounted for 40 percent of usage. Another example of an app that meets a specific need in a specific context would be the real-time reporting of study room usage so students can see ahead of time what rooms may be available before heading to the library.

FUTURE DIRECTIONS

Making sure that librarians get a chance to use mobile technology and apps is important in implementing mobile access to technology in a library. Training and support from technical staff is crucial. Tablet devices can also be used to facilitate many aspects of a librarian's daily work. Keeping up with new mobile technology and library apps can be a daunting task, and is best shared among several librarians. Getting familiar with mobile technology and using it in daily life will allow librarians to keep up with library users who are increasingly mobile-oriented, as well as stimulate finding new ways to offer creative and innovative services via mobile technology. Mobile technology allows us to reach beyond the physical library and provide services to many more clients than may ever actually visit our library spaces.

Mobile technology also allows libraries to provide new and innovative ways to access information. Pew Research Center recently reported on some innovative mobile apps and websites produced by libraries.[13] The Cuyahoga County Public Library created a mobile app that allows patrons to check out digital materials from their extensive digital library directly to their mobile devices, as well as provides access to the catalog and general library information. A GPS-aware locator allows patrons to find library branches near their location. Developers at the Goethe-Institut New York created an innovative app that uses augmented reality to allow exploration of German history in New York City. When a mobile device camera is pointed at a location, multimedia images are layered on top of the camera image showing what the building looked like in the past and providing historical information. The German Traces NYC app received the American Library Association 2013 Cutting Edge Library Service award. In the future, as mobile technology usage becomes ubiquitous in public areas and services, many more innovative ways of providing information to library users will become possible. Mobile access to library services will become the norm. Make sure your library is ready; start exploring providing mobile access to your libraries' services and collection now.

NOTES

1. Silipigni Connaway, Lynn, Donna M. Lanclos, and Erin M. Hood. "'I Always Stick With the First Thing That Comes Up on Google. . .': Where People Go for Information, What They Use, and Why." Educause Review, December 6, 2013. http://www.educause.edu/ero/article/i-always-stick-first-thing-comes-google-where-people-go-information-what-they-use-and-why (accessed May 30, 2014).

2. Fox, Zoe. "Mobile-App Use Increased 115% in 2013." Mashable, January 14, 2014. http://mashable.com/2014/01/14/mobile-app-use-2013/ (accessed May 30, 2014).

3. Perez, Sarah. "Mobile App Usage Increases In 2014, As Mobile Web Surfing Declines." TechCrunch, January 4, 2014. http://techcrunch.com/2014/04/01/mobile-app-usage-increases-in-2014-as-mobile-web-surfing-declines/ (accessed May 30, 2014).

4. Hickman, Martin. "Readers Are Now Buying More E-Books than Printed Books." *Independent*, August 6, 2012. http://www.independent.co.uk/arts-entertainment/books/news/readers-are-now-buying-more-ebooks-than-printed-books-8008827.html (accessed May 30, 2014).

5. Vollmer, Timothy. "There's an App for That! Libraries and Mobile Technology: An Introduction to Public Policy Considerations." American Library Association, June 2010. http://www.ala.org/offices/sites/ala.org.offices/files/content/oitp/publications/policybriefs/mobiledevices.pdf (accessed May 30, 2014).

6. mobiThinking. "Global Mobile Statistics 2014 Home: All the Latest Stats on Mobile Web, Apps, Marketing, Advertising, Subscribers, and Trends . . . " MobiForge, June 13, 2014. http://mobiforge.com/research-analysis/global-mobile-statistics-2014-home-all-latest-stats-mobile-web-apps-marketing-advertising-subscriber (accessed July 1, 2014).

7. AT&T. "The App Advantage." AT&T Small Business, 2013. http://www.att.com/Common/about_us/pdf/small_biz_app_advantage.pdf (accessed May 30, 2014).

8. Burrows, Peter. "Apple Infiltrates $3.8 Trillion Market With iPad: Tech." Bloomberg.com, February 1, 2012. http://www.bloomberg.com/news/2012-02-01/apple-invades-3-8t-workplace-market-with-ipad.html (accessed May 30, 2014).

9. Gruman, Galen. "The iPad Revolution Is Coming to a Hospital Near You." InfoWorld, January 25, 2013. http://www.infoworld.com/d/consumerization-of-it/the-ipad-revolution-coming-hospital-near-you-211472 (accessed May 30, 2014).

10. Lotts, Megan, and Stephanie Graves. "Using the iPad for Reference Services: Librarians Go Mobile." *College & Research Libraries News* 72, no. 4 (2011): 217–20.

11. Thompson, Sara Q. "Setting Up a Library iPad Program: Guidelines for Success." *College & Research Libraries News* 72, no. 4 (2011): 212–36.

12. Washburn, Bruce. "Library Mobile Applications: What Counts as Success?" *Information Outlook* 15, no. 1 (2011). http://www.oclc.org/research/publications/library/2011/washburn-io.pdf (accessed July 1, 2014).

13. Zickuhr, Kathryn. "Innovative Library Services." Pew Internet Libraries RSS, January 29, 2013. http://libraries.pewinternet.org/2013/01/29/innovative-library-services-in-the-wild/ (accessed July 1, 2014).

RESOURCES

Apple iPhone in Business: Profiles: http://www.apple.com/iphone/business/profiles/
Boopsie: http://www.boopsie.com/
Gallery of Mobile Apps and Sites: http://www.nlm.nih.gov/mobile/
German Traces NYC: http://www.germantracesnyc.org/
iMedical Apps: http://www.imedicalapps.com/
Lee, Angela, and Ann Whitney Gleason. "Tablet Mania: Exploring the Use of Tablet Computers in an Academic Health Sciences Library." *Journal of Hospital Librarianship* 12, no. 3 (2012): 281–87.

Peters, Thomas A., and Lori Bell. *The Handheld Library: Mobile Technology and the Librarian*. Santa Barbara, CA: ABC-CLIO, 2013.
Skyscape: http://www.skyscape.com/index/home.aspx

Chapter Six

Apps for Every Library

There is an app for just about everything, from business to education and from recreation to health and wellness. Apps are growing exponentially for all mobile operating systems, especially iOS and Android. Mobile users prefer apps for accessing their favorite social networking sites, playing their favorite games, and accessing cloud sites, because it is easier and faster to open an app that goes directly to what you need rather than browsing the Internet or even using bookmarked websites on your mobile device home screen. There are actually more mobile devices than personal computers sold each year and in 2012, Nielsen reported that more than half the time spent on a mobile phone is spent using apps.[1]

Marketing researchers have found that the use of branded apps increases interest in a product due to a high level of user engagement when using a mobile device.[2] Mobile users have strong emotional attachments to their devices, and informational apps that are useful in daily life are the most successful. Boston analytics and marketing company. Localytics, reported in 2011 that 26 percent of apps downloaded are only used once.[3] Before downloading or purchasing an app, it is a good idea to read the reviews on the iTunes or Google stores. There will be a link to the developer's website, which may contain screenshots or demos of the app to preview. You can also search the web for reviews and critiques of an app. When looking for a specific type of app, check for review articles on the best apps of that type to narrow down your choices. If the app is free, check to make sure it does not contain annoying ads, or send constant pop-up messages. Also, be sure to check the privacy policy for specific apps that may use location awareness. As with all websites, be sure you are downloading and installing an app that was created by a reputable source and does not contain malware, which can allow hackers to gain access to your personal data.

In this chapter we explore popular apps that are applicable to libraries, from general apps for business use, to apps created specifically for libraries, apps for access to library resources, and apps for medical information, as well as useful utilities. The apps are organized by category, with an explanation of that type of app's use in a library setting. Each of these apps have been used and reviewed by librarians and found to be stable and useful for research, education, or business pursuits. All of the apps come from an authoritative source, and most are free or inexpensive and do not contain ads. These apps have been tested for ease of use and functionality and are recommended for use by all audiences, although some that are harder to use have been included because of the high quality of the information contained.

BUSINESS AND PRODUCTIVITY

Business and productivity apps include those that allow note taking via keyboard, touch, or voice; mobile versions of web-conferencing software; apps for sharing documents; and mobile access to popular productivity software such as Microsoft Word, Excel, and PowerPoint, Apple iWork software, and Google Docs. This type of app enables writing, note-taking, research, and business transaction on the road, from home, or while on the bus or in an airplane without the need to carry a bulky laptop computer. In recent years, there has been an explosion of productivity apps for all mobile devices, enabling users to do work anywhere and anytime directly from their mobile devices. Popular software vendors have focused product development time and money on producing app versions of their productivity software to meet user demands. As mobile devices become more powerful, fully featured productivity software is becoming a reality for mobile apps. The following list contains information on some of the most popular apps for business and productivity on the market today:

- Adobe Connect Mobile: Host, attend, and present Adobe Connect meetings from your mobile device. Free for iOS, Android, BlackBerry.
- Awesome Note: Calendar, notes, planner, to-do lists, and more all in one highly graphical interface. Under $10 for iPad and iPhone, and free on Galaxy Note 8.
- CarbonFin Outliner: Create outlines quickly and share online. Under $10 for iOS.
- Dragon Dictation: Voice recognition and speech-to-text transcription for email, text messaging, and posting to Facebook and Twitter. Free for iPhone and iPad.
- Citrix ShareFile QuickEdit: Create and edit Microsoft Office–compatible documents on your iPad or iPhone. Free for iOS.

- Dropbox: Popular cloud application for sharing files of any kind between devices. Free basic account for iOS, Android, BlackBerry, Kindle Fire.
- Evernote: Popular note-taking cloud application, which allows syncing between devices. Free basic account for iOS, Android, BlackBerry, Windows Phone.
- Google Drive: Access Google Drive files from your mobile device. Free for Android, iOS.
- Keynote: Mobile version of the popular Apple presentation software. Also available is a separate Keynote remote app for controlling a presentation from your mobile device. Under $10 for iOS.
- Microsoft Office Mobile: View, create, and edit Microsoft Word, Excel, and PowerPoint documents on your mobile device. Free with eligible Office 365 subscription, for iOS, Android, Windows Phone.
- Pages: Mobile version of the popular Apple word processing software. Under $10 for iOS.
- PenSupremacy: Handwriting and sketching app for Android tablets. Under $10 for select newer Android tablets.
- Penultimate: Handwriting and sketching app for iPad; syncs with Evernote. Free basic account for iPad only.
- Simplenote: Simple note-taking app that syncs with the web for saving and sharing with others. Free for iOS, Android, Kindle.
- Speak It!: Text-to-speech app that reads aloud your emails, documents, and new articles. Under $10 for iOS.
- WebEx Mobile: Schedule, host, and attend WebEx meetings from your mobile device. Free for iOS, Android, Windows Phone, BlackBerry.

APPS FOR READING

Reading on mobile devices is complicated by the fact that e-books come in multiple formats. Proprietary e-books may also have digital right management (DRM) software attached to the file to control usage. The use of DRM can severely limit the way an e-book can be used and makes it difficult to use e-books for academic purposes that would otherwise be acceptable under academic fair use laws. Further complicating the e-book market is that multiple file formats are used by different vendors. Some of these are proprietary to the type of e-book reader used such as the Mobipocket format used by Amazon Kindles. There are converters for proprietary formats, but software and apps are now available for reading multiple file formats, which makes using e-books much easier. Each mobile reader has a basic reading interface with touch-enabled menus and page turning as well as additional features for bookmarking, table of contents, annotations, and other features. Some work with multiple file formats and some use cloud technology to allow syncing between multiple devices.

According to Digital Book World, the phenomenal growth in e-book sales is slowing.[4] Reading on the bus or airplane from an e-book reader is convenient, but consumer research indicates that many readers prefer physical books for recreational reading. Many people also prefer to purchase print books for nonfiction reading. The user experience when reading from an e-book reader is improving, but an interactive table of contents and indexing for easily moving to different sections of a book is still not a reality. At the University of Washington Health Sciences Library, we recently completed a survey of professional student technology preferences. Half of the users preferred reading print books rather than using e-books. Many of the e-book users owned a Kindle, but an equal percentage of readers used their tablets for reading e-books. Listed below are popular apps for reading e-books and listening to audiobooks:

- Audible: Free audiobook reader app from Amazon. Free for iOS, Android, Windows Phone, Kindle.
- Bluefire Reader: Read ePub, PDF, and Adobe DRM e-books on your mobile device. Free for iOS and Android.
- Browzine: Browse open-access and institutional subscription journals (for participating universities or colleges) from your mobile device. Free for iPad, Android, and Kindle Fire HD tablets.
- Flipboard: Add news from the web to your personal magazine collection. Free for iOS, Android, Windows Phone, BlackBerry, Kindle Fire, and Nook.
- Goodreads: Mobile app for the popular book discussion and recommendation site. Free for iOS, Android, Nook.
- Google Play books: Choose from millions of books (some free, some fee-based) on Google Play. Free app for iOS and Android.
- iAnnotate PDF: Annotate PDFs, documents, and images on your tablet device and save to Dropbox or Google Drive. Under $10 for iPad, free for Android.
- iBooks: Free e-book reading app with access to online books for purchase. Free for iOS.
- Kindle: Amazon's free e-book reading app. Download books from the Amazon store and read from multiple devices. Free for iOS and Android.
- OverDrive Media Console: Download e-books, audiobooks, and videos from public libraries to your mobile device. Free for iOS, Android, Kindle, Nook, Windows Phone.

RESEARCH APPS

Many vendors have created mobile web formats for their online databases, and some are beginning to release fully featured mobile apps for their customers. If a database vendor has a website redirect in place, optimized for small-format mobile screens, it is a best practice to also allow a link to connect to the full website as some features may not be available on the mobile interface. Tablet computers have large enough screens so that database search websites display pretty well without a specially formatted mobile display. Mobile apps for access to research databases are especially useful for quick reference lookups, but are often limited compared to the full website. Science vendors have been early adopters of mobile apps, but a few other disciplines are also creating mobile interfaces. These apps are essential to librarians doing reference outreach in a clinical or educational setting where a desktop computer may not be available. Mobile apps for connecting to research databases allow a librarian or researcher to quickly access a popular database for quick reference and download research materials. The iBook app is helpful for reading and saving downloaded research papers directly on your mobile device. For more information, see the list of reading apps included in this chapter. The list below contains some of the more popular mobile apps for academic research purposes:

- arXiv mobile: Access Cornell's large arXiv repository of scientific e-papers. Free for iOS, and under $10 for Android.
- The Chemical Touch: Lite Edition: Easy-to-use periodic table of information for all the elements. Free for iOS, Android, and Windows Phone.
- Dictionary.com: Full-featured dictionary app with voice search. Free for iOS, Android, Windows Phone, BlackBerry, Kindle Fire, and Nook.
- Free Books: 23,469 Classics To Go: Browse a large collection of classic books with a full-featured reader including notes, highlighting, bookmarks, and dictionary support. Free for iOS, Android, and Windows Phone.
- HeinOnline app: Popular legal research database access from your mobile device. Free for iOS but requires institutional HeinOnline account/authentication.
- History: Maps of the World: Access to a collection of interactive historical maps showing geopolitical changes over time. Free for iOS.
- iSSRN: Access research in the fields of humanities and social sciences from the Social Science Research Network (SSRN). Free for iOS.
- Planets: Popular 3D guide to the solar system. Free for iOS and Android.
- PLoS Reader: Access the Public Library of Science journal articles from your mobile device. Free for iOS and Android.

- PocketJustice: U.S. Supreme Court justice information, records, and case information in a mobile format. Under $1 for iOS and Android.
- PubGet: Download full-text scientific articles available through your institutional affiliation from PubMed and other top databases. Free app for iPad.
- PubMed On Tap: Search PubMed and access full text through your institutional affiliation/authentication. Under $10 for iOS.
- REFScan: Scan a DOI to access the bibliographic information and times cited from Web of Knowledge and save to EndNote Web. Free for iOS.
- Reuters News Pro: Access Reuters' up-to-date reporting and market data from your mobile device. Free for iOS and Android.
- ScienceDirect Mobile: Search for articles and access full text through your institutional subscription. Free or paid version for iPhone and Android.
- TED: Access a library of TED Talks from your mobile device. Free for iOS.
- WolframAlpha: Access the powerful computational and knowledge search engine from your mobile device. Under $10 for iOS, Android, Kindle Fire, and Nook.

CITATION MANAGEMENT APPS

Most of the popular citation management tools have a mobile app available to access your reference libraries on the go. Most of the citation management tools are free, and allow adding resources using the app and syncing with the desktop or web version of the software. Even though these are scaled-down mobile versions of fully featured citation management tools, most allow research organization and annotation as well as access to your research libraries. Another nice feature is the ability to use your mobile device camera to scan a bar code to automatically upload a book reference to your reference library. This is extremely useful when doing library research with physical materials, allowing direct uploading of citations using your mobile device. All of the mobile versions of popular citation management software include multiple formats for generating citations. Some also allow automatic searching of online databases such as PubMed in order to download references. Citation management apps include:

- Droideley: Sync and download documents from your Mendeley library from your Android phone or tablet. Free for Android.
- EasyBib: Scan a bar code or search for a book and generate citations in multiple formats. Free for iOS and Android.
- Endnote: Access your Endnote library and annotate PDFs from your mobile device. Under $10 for the iPad.

- Mendeley: Access your Mendeley library to read and annotate research papers from your iOS device. Free for iOS.
- Papers: Search for articles using the built-in search engines and organize your research from your iOS device. Under $20 for iOS.
- Zotpad: Access your Zotero libraries from your mobile device. Free for iOS.

MEDICAL APPS

The medical community was an early adopter of mobile technology and so there are many medical-themed apps available. Most health sciences resource vendors now have mobile-optimized search interfaces, and many are releasing mobile apps as well. Busy health care workers have embraced mobile technology to enable evidence-based practice at the bedside. Many health sciences libraries provide lists of apps for access to science and clinical resources and tools. Categories of medical apps include clinical diagnostic tools for tests and codes, clinical decision support tools, drug resources, clinical calculators, and research database apps, as well as patient education apps and personal wellness apps, which is a fast-growing market. Apps that facilitate management of diseases for patient use are also a growing market, although apps for easy access to patient records are still a thing of the future. Modern Healthcare recently reported that only 54 percent of the medical apps available at the time from the iTunes store were truly useful in a clinical setting.[5] Librarians can help sift through the growing list of medical apps to provide curated lists of the best and most authoritative apps available. Using mobile devices in a health care setting can be complicated by federal Health Insurance Portability and Accountability Act (HIPAA) rules, but using standard security measures and not downloading patient information onto your device should prevent any problems. The list below contains a few of the more popular and well-tested medical apps available for research and clinical use:

- AHRQ ePSS: The Electronic Preventive Services Selector (ePSS) allows access to the U.S. Preventive Services Task Force recommendations. Free for iOS, Android, Windows Phone, BlackBerry, and Palm.
- ARUP Consult: Physician guide to laboratory test selection with links to PubMed articles online. Free for iOS.
- Calculate by QxMD: Clinical calculator and decision support tool. Free for iOS, Android, and BlackBerry.
- Diagnosaurus: Database of differential diagnoses; search by disease, symptom, or organ system. Under $10 for iOS and Android, free for BlackBerry, Windows Mobile, and Palm.

- Draw MD: Allows visual communication of anatomy, conditions, procedures, and medical concept with patients. Free for iPad.
- Dynamed: Mobile access to the Dynamed database through your institutional subscription. Free for iOS and Android.
- Epocrates Rx: Access to drug information and interactions as well as pill identification and formulary. Free for iOS and Android.
- Medscape: Evidence-based disease and drug reference, as well as medical news by specialty. Free for iOS, Android, and Kindle Fire.
- Micromedex: Free drug reference through your institutional subscription, or purchase an individual account for a small fee. Free for iOS.
- PubMed for Handhelds: Search PubMed from your iPhone or iPad. Free for iOS.
- Radiology 2.0: One Night in the ED: Interactive radiology educational app. Free for iOS.
- Skyscape: Free medical library and listing of available medical apps. Free for iOS and Android.
- STAT ICD-9 Lite: Medical diagnosis coding reference, search by disease classification. Free for iOS.
- VisualDX: Free mobile app with access to the visual differential builder for dermatology diagnosis. Free for iOS and Android.

UTILITIES

Utilities apps are one of the largest categories of apps available on the iTunes and Google Play markets. Well-chosen utility apps can make your mobile device even more useful in daily life. Utilities apps range from tools for using your mobile device to tools for connecting to other devices and peripheral equipment, as well as popular tools for quick reference information such as calculators, maps, and language translators. Librarians can help mobile device users by sifting through the many available apps to find the most useful utility apps from the most authoritative sources. There are many utilitarian apps that allow quick access to information, or provide tools to make education and business pursuits easier and more mobile. The following are some of the more popular apps that may be useful for libraries:

- Amazon: Free app to access your online Amazon account, browse and purchase books, videos, and more. Free for iOS, Android, and Windows Phone.
- Babylon: Free translator for your mobile device. Free for iOS, Android, Windows Phone, BlackBerry, and Kindle.
- Calc Pro: Multiple calculators in one app. Under $10 for iOS, Android, and Windows Phone.

- Doc To Go: View, edit, and create Microsoft Office documents from your mobile device. Under $10 for iOS an Android.
- Google Maps: Popular navigation app from Google. Free for iOS and Android.
- JotNot Scanner: Use your phone's camera to scan documents and email or upload to cloud apps. Free for iOS.
- LogMeIn: Access your desktop computer remotely from your mobile device. Free for iOS and Android.
- MindNode: Create detailed mind maps on your mobile device and save to the cloud. Under $10 for iOS.
- Popplet: Easy mind mapping for iPad. Free lite version, or under $10 for iOS.
- Print n Share: Print to a wide variety of networks on wireless printers from your iOS device. Under $10 for iOS and Android.
- QR Reader: Simple and free QR code reader and creator. Free for iOS and Android.
- RealCalc Scientific Calculator: Popular, visually realistic scientific calculator for Android devices. Free for Android.
- Scan2PDF Mobile: Create multipage PDFs using your device's camera. Free for Android.
- SyncPad: Cloud-based whiteboard that allows collaborative drawings and sharing of images and pdfs. Free for iOS.
- Wyse Pocket Cloud: Remotely connect to your desktop computer from your mobile device. Free for iOS and Android, or purchase Pro version for under $20.

NEWS APPS

News apps are the third most popular types of app downloaded in the United States after social networking and shopping apps, according to Nielsen, and one-third of mobile device users surveyed had downloaded a news app in the past month.[6] Furthermore, 77 percent of tablet users surveyed in another study reported reading news through their apps at least weekly. Newspaper reading is rapidly transitioning from paper to digital formats, and many top news sources are producing apps to access their articles. Many of these news sources provide free apps to access their content, although the apps often include ads. There are many news apps to choose from, but here are a few top-rated ones:

- AP Mobile: Associated Press mobile app. Free for iOS and Android.
- BBC News: Get the latest global news from the BBC. Free for iOS and Android.

- CNN: Watch live CNN programming on your mobile device. Free for iOS, Android, Windows Phone, BlackBerry, Kindle Fire, and Nook.
- ESPN ScoreCenter: Sports scores, news, and video highlights for mobile devices. Free for iOS and Android.
- NPR News: Follow local and national news and listen to your favorite NPR stations. Free for iOS and Android.
- NYTimes: Browse free *New York Times* news articles, blogs, and videos from the app, but requires a subscription to access all materials. Free for iOS, Android, Window Phone, BlackBerry, and Kindle Fire.

SOCIAL NETWORKING APPS

Social networking apps are the most popular category of apps downloaded and accessed from all devices. If your library is using popular social networking applications for marketing your services and connecting with library users, you will want to be familiar with the mobile versions. In addition, the use of social networking apps in education is increasingly popular. The use of social networking sites, especially on mobile devices, is integrated into the daily lives of millions of people. Using social networking tools, libraries and educators can reach out to users and students by providing an interactive learning environment that is accessible anywhere. Blogging software is increasing in popularity for publishing online as well as communicating in the academic world. All social networking apps require a free account to be set up when downloading and installing the app. Security measures should always be followed, and care taken to protect personal information when using social networking apps. Popular apps include:

- Facebook: Connect to Facebook from your mobile device. Free for iOS and Android.
- Google+: Connect to your circles and hangouts and make video calls with the Google+ app. Free for iOS and Android.
- LinkedIn: Connect to your business contacts anywhere. Free for iOS, Android, Windows Phone, and BlackBerry.
- Skype: Face-to-face video calls and texting from your mobile device. Free for iOS, Android, Windows Phone, and BlackBerry.
- Twitter: Get news and updates through your Twitter feeds on your mobile device. Free for iOS, Android, Windows Phone, and BlackBerry.
- WordPress: Write and edit blog posts as well as manage comments on a mobile device. Free for iOS, Android, and BlackBerry.

ORGANIZING APPS

The more apps you add to your device, the more screens you will have to scroll through, making it hard to find what you are looking for. One way to organize your apps is to put similar apps on the same screen. On an iPad or iPhone, pressing on an app and holding will make all the icons shake, and then you can press and move your apps around the screen or between screens. When you fill up a screen with apps, it will create another screen, and so on up to 15 screens with the current version of iOS. Most Android tablets or phones come with five screens. To move apps, press and hold, and then drag the icon to the left or right to move to another screen. You can also organize apps in folders. On an iOS device, press and hold on an app until they all start shaking, then just drag one icon on top of another one to automatically create a folder. You can name the folder to identify the type of apps contained. There can be up to 20 apps in a folder. On an Android device you can also create folders by dragging one app on top of another. Older versions may have other methods of creating folders. The number of apps that can be put in a folder differs depending on the version of Android and the model of your phone.

Keeping up with new apps can be difficult, since new ones are constantly being released. You can browse the Google Play Store or the iTunes App Store for new and featured apps, or search under the categories provided. There are also lists of apps for specific subjects on websites such as iMedicalApps.com. There are many websites that review apps, and online magazines such as PC Magazine often provide this service. There are even apps for recommending apps, such as AppGrooves. Keeping apps updated can also be time consuming, as updates are released all the time. It's important to run updates often to install bug fixes and new features, but check to make sure the update will work with your model of phone and the apps you use frequently.

Keeping lists of apps on your library's website or in LibGuides for subject area specialties is a helpful service that libraries can easily provide. When reviewing and evaluating apps, it is helpful to use a checklist of criteria for selection. Is the app from an authoritative source? Are there ads included, and is the information reliable? Is the content appropriate for the audience you are selecting for? What is the cost of the app, or is it free? Another consideration that may be important is whether or not the apps can be used offline when a network connection is not available. It is clear that apps can be used in many ways in libraries. Apps that access library databases can be demonstrated in library orientation and instruction sessions. More about using apps for teaching and learning will be discussed in chapter 8. Using apps for outreach and communication purposes in libraries will be explored in chapter 9.

NOTES

1. "State of the Media: The Social Media Report 2012." Nielsen Reports and Insights, December 4, 2012. http://www.nielsen.com/us/en/reports/2012/state-of-the-media-the-social-media-report-2012.html (accessed June 26, 2014).

2. Bellman, Steven, Robert F. Potter, Shiree Treleaven-Hassard, Jennifer A. Robinson, and Duane Varan. "The Effectiveness of Branded Mobile Phone Apps." *Journal of Interactive Marketing* 25, no. 4 (2011): 191–200.

3. "Mobile App Analytics Show 26% of App Downloads Used One Time." Localytics, January 31, 2011. http://www.localytics.com/blog/2011/first-impressions-26-percent-of-apps-downloaded-used-just-once/ (accessed June 26, 2014).

4. Greenfield, Jeremy. "Ebooks Up Modestly in First Quarter 2013." Digital Book World, July 16, 2013. http://www.digitalbookworld.com/2013/ebooks-up-modestly-in-first-quarter-2013/ (accessed June 28, 2014).

5. Conn, Joseph. "Mobile Medical Apps Are Becoming Mainstream for Doctors and Patients." Modern Healthcare, December 14, 2013. http://www.modernhealthcare.com/article/20131214/MAGAZINE/312149983 (accessed June 28, 2014).

6. Sonderman, Jeff. "Nielsen: One-third of Mobile Users Downloaded News Apps in Past Month." Poynter, January 9, 2012. http://www.poynter.org/latest-news/media-lab/mobile-media/158833/nielsen-one-third-of-mobile-users-downloaded-news-apps-in-past-month/ (accessed June 28, 2014).

RESOURCES

8 Apps That Make Academic Research Easier: http://www.maclife.com/article/gallery/8_apps_make_academic_research_easier

25 Apps You'll Need to Survive College: http://mashable.com/2013/08/08/apps-for-college/

50 Must-Download Apps for Lifelong Learners: http://www.edudemic.com/50-must-download-apps-for-lifelong-learners/

61 Educational Apps for the 21st Century Student: http://www.teachthought.com/apps-2/61-educational-apps-for-the-21st-century-student/

Apps for Academics: Mobile Web Sites and Apps (Research Guides at MIT Libraries): http://libguides.mit.edu/apps

Apps for Book Lovers: http://appadvice.com/applists/show/apps-for-book-lovers

Comparison of e-book Formats: Wikipedia, the Free Encyclopedia:http://en.wikipedia.org/wiki/Comparison_of_e-book_formats

Digital Rights Management: Wikipedia, the Free Encyclopedia: http://en.wikipedia.org/wiki/Digital_rights_management

Free Apps for Learning (WVU Libraries): https://www.libraries.wvu.edu/libraries/evansdale/forms/apps.pdf

Six Steps to Decluttering Your Smartphone's Apps: http://www.nytimes.com/2012/06/14/technology/personaltech/six-steps-to-decluttering-your-smartphones-apps.html?_r=0

Top 10 Free iPad Medical Apps for Health Care Providers: http://www.imedicalapps.com/2012/06/top-free-ipad-medical-apps/

Chapter Seven

Best Practices for Using Mobile Technology in Libraries

Best practices are important guidelines to follow to help ensure successful mobile initiatives and the best user experience. Although these guidelines are important, their utility needs to be weighed against each unique library situation and community need. Different communities and institutions can have very different needs depending on the primary users of the library services and resources, as well as characteristics of the community that the library serves such as existing resources, location, and demographic information. These guidelines for using mobile technology in libraries are meant to help with the creation and implementation of mobile programs, but each library should always consider the actual needs of the population served as well as align with existing library policies and procedures.

Before we consider best practices for specific types of mobile technology projects that might be implemented in libraries, here are some general best practices for any mobile project:

- Know your audience and the services they need and want
- Assess user needs and make sure your mobile project fulfills an important community need
- Consider starting with a pilot project or beta app instead of jumping into a full-blown project right away
- Conduct usability studies and focus groups to assess a mobile product or project, or conduct an online survey if your service is for remote users
- Make sure you have trained technical support available, whether it be library support staff or an outside company
- Start with a project that fits your library's expertise as well as budgetary limitations

- Use web analytics to discover what mobile devices access your existing websites, in order to target your program and adjust as needs change

A mobile project such as building an app or optimizing a website is no different from the process used when creating a traditional library website, and established best practices for building good library websites should also be followed. Library best practices for creating projects and services should also be considered when beginning a mobile technology project for your library. It is also very important to explore your institution's policies and procedures to make sure your project aligns with existing practices to avoid possible future conflicts that could impact the success of your library mobile project.

BEST PRACTICES FOR MOBILE-FRIENDLY WEBSITES

The Google Developers site at developers.google.com has many helpful resources and tips for building mobile-friendly websites.[1] There are basically three options for setting up a mobile-friendly website: responsive design, which uses the same website for all devices, but uses CSS to adjust the content to fit smaller screens; dynamic web pages, which display different websites depending on the type and size of the device that is accessing the content; and a separate mobile site with a separate URL for mobile devices, to access content formatted for smaller screens. Google suggests using the first option, responsive design. This method avoids redirects to different websites and works for all types of devices, including ones that may be new on the market. You might consider not creating a separate display for tablet computers, especially the larger-screen models, since the display is close to the size of a desktop monitor and most websites will display normally.

Besides offering specific programming details on how to create different types of mobile-friendly websites, the Google Developers website also offers tips for avoiding common mistakes. If your website is designed to redirect users to different websites depending on the type of mobile device that is detected, make sure that all content redirects to the correct site and that you have appropriate sites set up for the different device types that may be accessing your site. You must also ensure that each redirect from the original website redirects to the correct device type. These common errors are one of the reasons Google recommends using responsive web design instead of redirects. Another common error is using video formats such as Flash that may not display correctly on mobile devices. If you want to use video on your website, the best practice is to use HTML5 video tags, provide a list of multiple formats, and include .mp4, which will play on Apple devices. Two other common formats for mobile devices are .ogg and .webm. It is also very important to consider the page load speed of a mobile web page, since mobile networks are usually much slower than the connections available to a desktop computer. Calls to JavaScript, and external CSS, which is usually loaded in the header section of a website's HTML, can significantly slow down page loading on a mobile device to the point where a user may give up on the website. Avoiding JavaScript and external CSS is helpful, but if needed these can be programmed to be loaded after a delay so that the initial webpage renders faster.

The World Wide Web Consortium (W3C), an international organization that works on open web standards, has created a lengthy document detailing specific guidelines for delivering web content to mobile devices.[2] This document introduces the concept of "one web," which is the idea that website content and services should be accessible to all users from any device. The W3C's Mobile Web Best Practices document outlines the specifications for a minimum default delivery context that should work on most devices. This specification includes a minimum screen size of 120 pixels, using XHTML Basic markup language and UTF-8 character encoding, formatting images as JPEG or GIF 89a, and refraining from using client side scripting. Establishing the context in which your website is accessed will help you determine what content you need to optimize for mobile device access. For example, some website content may be more commonly accessed by mobile devices, such as information about library hours or study room reservations. Other content may be more commonly accessed from a desktop computer. According to the W3C, users should always be allowed a choice of accessing the mobile-optimized website or the full website. In addition, the W3C document outlines a series of specific best practice, mobile web design principles for navigation, layout, page definition, and user input.

The "mobile-first" strategy addresses the problem of designing a website that works for all of the many operating systems and models of mobile devices currently in use, including those that may be developed in the near future. A mobile-first strategy starts with a mobile-responsive website that works for all sizes of screens and types of devices. Then developers can customize the design for popular devices, and later add optimization for new devices as they appear on the market. According to the Mobile Web Best Practices website, the first step in building a mobile-optimized website is to step back from design work and consider your content.[3] Consider what content is most important for your website's users, and why.

Exploring the logical structure of your existing content can help you realize what information is most important to present to users, and what content logically goes together. Structuring also helps you create a strategy to make your website content more cohesive and easy to browse. It is recommended that you start with a website design that works well for all mobile web browsers, not just the popular ones, which is the worst-case scenario.

Then, the design process can scale up through different sized screens, arriving at a version that also works on a desktop computer. That way, you won't be leaving users with unusual or outdated devices behind. Optimizing your website for specific devices will depend on the devices most commonly used to access your website; web analytics products such as Google Analytics can help you identify which devices and browsers are accessing your website and what content they are most commonly viewing.

BEST PRACTICES FOR SELECTING APPS

When selecting apps to recommend to your library users or for including on mobile devices that will be checked out, there are many variables to consider. The app must first and foremost be of relevance to your target audience, whether the audience is senior citizens or second graders. Deciding if a library mobile app program will have a specific discipline or interest area focus will also be important when choosing apps. The cost of apps may also be an important deciding factor since budgets are often limited. Fortunately, there are a lot of quality apps available for free or at low cost. Measurements of the quality of a mobile app are somewhat hard to define. Some quality measurements might include the level of interactivity provided, the educational level needed to understand the content, ease of use and navigation, as well as the amount of content provided. Fortunately, there are many websites that evaluate and list apps for various subjects. When looking for educational apps, there are some great websites available to educators, such as the Granville Elementary School best educational apps site. Many higher education libraries have created LibGuides for the best apps in particular subject areas. These can easily be found by doing a Google search. Some excellent higher education examples include the MIT Libraries' listing of apps for academics at http://libguides.mit.edu/apps and the Yale Medical Library listing of medicine-related mobile apps at http://library.medicine.yale.edu/services/computing/mobile_apps.

The Mobile Technologies in Libraries website and blog is a great source of information to guide mobile technology programs.[4] This website has published a series of guides called "Pathways to Best Practices." In the guide on providing mobile access to resources, lessons learned include gathering information on devices used by your institution's users and doing research and assessment to find out how your specific audience is using their mobile devices in order to target your mobile programs. Continue to assess and gather feedback to constantly improve your programs and react to changing needs. Do usability testing to involve users in your programs and gather important feedback to enhance user interfaces. Also, when providing access to vendor resources, be sure to test each interface and provide instructions for access and authentication if needed. Database vendors' mobile interfaces are extremely varied and some emphasize secure access over functionality, making it sometimes difficult and tedious to access their resources. Restrictions on use and periodic reverification of access can cause users to abandon the use of valuable data resources due to the difficulties built in to access methods. Librarians can provide valuable feedback to vendors and lobby for secure but simplified access methods.

BEST PRACTICES FOR BUILDING APPS

The Apple programming guides contain a wealth of information for would-be iOS app creators.[5] Once you have the basic idea for an app, it's best to do some work on paper rather than jumping right into programming code. The Apple programming guide recommends starting with a full product description, including what you want your app to do, what your users will want from your app, and all the features you intend to include. Make sure your ideas about app features match what your users may want. Become familiar with all the capabilities of iOS by reading the pertinent materials on the developer website so that you get a better idea of how to best use iOS to achieve the goals you have set for your app project. Finally, drawing out your design ideas on paper will help you get a better idea of what you want the user interface to look like. Some important questions to answer before getting started include whether to create a universal iOS app or one specific to a device, and what user information you will need to collect and store in order for your app to work correctly. Best practices mandate that you collect the minimum amount of user data needed, and that you are transparent about what you are collecting and how it will be used. Protecting app user privacy is an important consideration when dealing with such a personalized device as a tablet or smartphone.

Since mobile devices are highly personalized, it is no surprise that mobile apps access and use highly personal information including demographics, financial information, passwords, and consumer and health information. The Center for Democracy and Technology (CDT) provides an in-depth online guide to protecting user privacy for developers.[6] The recommendations include being fully transparent and disclosing any and all data that may be collected, who it is shared with, and what it is used for. A privacy policy should be prominently displayed and users notified when the policy changes. Users should be allowed the option of not saving or sharing data if desired. App developers should make sure that all saved data is automatically deleted when a user's account is discontinued. Always use up-to-date software and development tools to ensure that user data is fully protected from current threats. Use security measures such as encryption, de-identification, and user authentication if sensitive data is collected. Be aware that for all applications for child audiences 12 and under, you must obtain parental consent before collecting any personal information, including names. Only collect the minimum amount of information needed for the functionality of your mobile application.

With the release of iOS 7, Apple published an online guide to best practices for building user interfaces that work well.[7] The iOS Human Interface Guidelines should be reviewed by anyone considering building an app, and many of the guidelines also apply to websites. First and foremost, Apple recommends that any user interface should put content and functionality first. Complex layouts and extra details or decoration may detract from the content that is displayed. Using negative or white space on the screen, no matter what the size, will help to highlight content and clarify functionality. Simplifying the colors used in an interface can also help clarify content, and the colors used throughout an app should be consistent. Similarly, using the system fonts rather than custom ones that may not be as legible or easy to read is recommended. Apple's guidelines for the user experience when starting and stopping apps include starting instantly to engage the user, and be prepared for the user to stop at any time. Starting instantly may mean that you delay gathering information from a user or requiring a login until the user has become engaged with your content. Since a mobile device user can switch to another application at any time, make sure that if the user decides to leave your app that no data will be lost and that they will return to the same place they left when they return to the app.

TIP: Put content and functionality first when designing for mobile devices. Don't be afraid to use negative space; decorative elements can be distracting and confuse users.

Navigation is an important element of any online content, and apps are no exception. Apple's iOS Human Interface Guidelines include several suggestions for ensuring that navigation is smooth and seamless. Navigation elements should work well in the background without being noticed by the user. If a user notices navigation, that usually means that something isn't working quite right and is not intuitive. There are three main types of navigation used in apps: hierarchical, flat, and content-driven. A hierarchical structure is characterized by presenting one choice per screen, with one question leading to another until the choices end. Users must then retrace their steps or start over from the beginning. A navigation bar can be provided to make this type of navigation easier. A good example of this kind of structure is the settings app on an iPhone. Flat navigation is characterized by a main screen, which presents multiple choices. With this kind of navigation, users can navigate directly between categories from the main screen. Tabbed browsing is a good way to provide flat navigation. An example of this kind of navigation is Apple's App Store. Content-driven navigation allows different content to be displayed depending on user choices. This kind of navigation is used for interactive apps such as iBooks or game apps. Some apps may use multiple types of navigation depending on the type of content being presented. In general, it is good practice to only provide one path to each screen in an app, no matter which type of navigation you use.

There are some common mistakes many app designers make that you should avoid. Branding your app with your colors and logo can be an important consideration, but do not overwhelm your content in the process of promoting your brand. If using custom colors, make sure they work well together on the screen. Color contrast is also important for readability. As a rule, there should be a contrast of at least 50 percent between colors when used together, such as colored buttons on a colored navigation bar. Color should not distract the user. Be aware that some colors will evoke different meanings and emotions in users. Another important consideration for apps as well as web content is avoiding the use of jargon and making your text understandable to the intended audience. Use a journalistic tone when presenting content; not too formal, but not too friendly either. Navigation labels should be short or use well-known icons to indicate actions. Having to stop to read long labels will interrupt the flow of an app. Make sure that all spelling and wording are correct, and consider using an impartial editor to provide feedback on the presentation of your content; this will ensure that it is presented in the way you intended.

Whether designing an app or mobile-friendly website, make sure your design does not exclude those with disabilities. A well-designed website or app is accessible for all users. Making an application accessible involves adding options for alternative modes of presenting content. The Apple Developer website has very useful guidelines for developing accessible applications.[8] To ensure accessibility for the visually impaired, make sure that color is not the only cue to a meaning or action so that your app will be accessible to color-blind users. Consider adding audio communication along with visual and textual instructions and feedback. Make sure images and video have text options so that screen readers will be able to access this content. To accommodate hearing impairments, make sure that all audio content is marked with a visual cue and provide a text-based option. Be aware that those with fine motor impairments may not be able to use a mouse or touch input device efficiently. Providing keyboard options for all navigation will make an application more accessible. An important consideration for users of assistive technology is to always design a clear way to exit an application or back out of navigation paths, which is good design practice anyway.

BEST PRACTICES FOR LIBRARY MOBILE PROGRAMS

The Association of College and Research Libraries' "Guidelines for Instruction Programs in Academic Libraries," revised in 2011,[9] includes a list of modes of instruction and technology tools needed for effective library instruction programs. These include:

- Group instruction in library or campus classrooms
- Web tutorials or web-based instruction
- Asynchronous modes of instruction (email, social media)
- Synchronous modes of instruction (chat, audio/video/web conferencing)
- Course management software
- Hybrid/distributed learning/distance learning, employing combinations of these methods

As we can see from the many examples provided in this book, many of these modes of instruction, such as web-based and asynchronous instruction, the use of a course management system, and distance learning methods, lend themselves well to the use of mobile technology. With the rapid growth in the use of mobile technology, libraries would be wise to consider adapting library instruction to include the use of mobile devices.

When developing a library mobile program of any kind, it is important to think strategically and develop a mobile strategy. Your college or institution may already have a mobile strategy in place to help guide library initiatives. The Educause publication, "Developing a Campus Mobile Strategy," reminds us that not all institutions are the same, so it is important to design your institution's mobile strategy with your unique user needs in mind.[10] All potential audiences of a mobile program should be considered. When creating a mobile strategy, consider what level of program must be offered. A good starting point is to offer access to real-time data that users need. More advanced programs may involve providing personalized interfaces and login access to database information. A data plan and possibly a data dictionary may also be needed to guide development of mobile applications that must access your institution's data in a consistent way. Also, since mobile technology changes so rapidly, a well thought out mobile strategy can help guide institutional directions over time and through changing hardware and mobile service offerings. It is also important to consider what technology support is necessary when planning for mobile technology programs, so that additional support resources can be found if needed. Budgetary considerations, such as identifying available program funding or grant opportunities are also important when planning what initiatives your institution should pursue at any point in time. Sustainability of a mobile technology program over time is another important consideration.

Guidelines and best practices for the use of mobile technology in libraries are still under development as the library profession slowly adopts this new technology into their practices and programs. As the use of mobile technology grows, much work still needs to be done to produce standards for libraries to follow when adopting this new technology to ensure successful, effective programs. More research is also needed to create standards for teaching and learning with mobile technology. Assessment of existing practices will lead to improvements in services over time. Standards relating to the most effective presentation of information for use with mobile devices, as well as standards for effective education programs using mobile devices, are also sorely needed. Research in this area would be extremely helpful in advancing teaching and learning with mobile technology.

NOTES

1. Google. "Building Smartphone-Optimized Websites." Google Developers, n.d. https://developers.google.com/webmasters/smartphone-sites/ (accessed August 15, 2014).

2. "Mobile Web Best Practices 1.0." W3C, July 29, 2008. http://www.w3.org/TR/mobile-bp/ (accessed August 16, 2014).

3. Frost, Brad. "Implement a Mobile-First Strategy." Mobile Web Best Practices, n.d. http://mobilewebbestpractices.com/strategy/implement-a-mobile-first-strategy/ (accessed August 10, 2014).

4. Alcock, Jo. "Pathways to Best Practice Guides." Mobile Technologies in Libraries RSS, August 23, 2012. http://mlibraries.jiscinvolve.org/wp/pathways-to-best-practice-guides/ (accessed August 19, 2014).

5. "iOS App Programming Guide: App Design Basics." Apple Inc., n.d. https://developer.apple.com/library/ios/documentation/iPhone/Conceptual/iPhoneOSProgrammingGuide/AppDesignBasics/AppDesignBasics.html (accessed August 15, 2014).

6. Future of Privacy Forum, Center for Democracy and Technology. "Best Practices for Mobile Application Developers." CDT.org, 2012. https://www.cdt.org/files/pdfs/Best-Practices-Mobile-App-Developers.pdf (accessed August 18, 2014).

7. "iOS Human Interface Guidelines: Designing for iOS 7." Apple Inc., n.d. https://developer.apple.com/library/ios/documentation/UserExperience/Conceptual/MobileHIG/ (accessed August 16, 2014).

8. "Accessibility Overview for OS X: Developing an Accessible OS X Application." Apple Inc., 2014. https://developer.apple.com/library/mac/documentation/Accessibility/Conceptual/AccessibilityMacOSX/OSXAXDeveloping/OSXAXDeveloping.html (accessed August 18, 2014).

9. ACRL Instruction Section. "Guidelines for Instruction Programs in Academic Libraries." *College & Research Libraries News*, ACRL, October 2011. http://crln.acrl.org/content/73/4/207.full.pdf (accessed August 18, 2014).

10. "Developing a Campus Mobile Strategy: Guidelines, Tools and Best Practices." Educause, January 2013. http://net.educause.edu/ir/library/pdf/ACTI1303.pdf (accessed August 18, 2014).

Chapter Eight

Using Mobile Technology in Education

The New Media Consortium's 2012 Horizon Report, which forecasts the trends in technology each year, focused almost exclusively on mobile learning.[1] Three of the top six key trends identified related to online learning. The top trend was "people expect to be able to work, learn, and study whenever and wherever they want to." The fourth trend was "the abundance of resources and relationships made easily accessible via the Internet is increasingly challenging us to revisit our roles as educators," and the fifth trend stated that "education paradigms are shifting to include online learning, hybrid learning, and collaborative models." We are transitioning to a mobile society and education is no exception, although the traditional educational model of an expert or "sage on the stage" lecturing to a classroom of passive learners is slow to give way to a more modern model where the instructor is more of a "guide on the side." This second model much more readily lends itself to online and mobile learning.

Online learning technology allows students to use mobile devices to access learning modules, which facilitates true interactive learning. Online learning can be highly individualized and occurs in the user's environment where the student can be comfortable and work at the pace needed for optimum knowledge building. Individualized learning methods allow the user to spend more or less time on learning modules as needed. Online learning follows the constructivist model of education, and provides opportunities for more active learning where the student "constructs" and internalizes knowledge. The teacher is more of a facilitator, and students are responsible for creating their own knowledge. This type of learning lends itself to informal, real-world learning activities such as problem-based learning, which also works well in a mobile setting. Breaking educational topics up into separate online learning modules works well to provide just-in-time information, which can be accessed from mobile devices anytime, anywhere, at the point of need.

BACKGROUND ON ONLINE LEARNING

Online education has been increasing in popularity as students become more and more immersed in personal technology. Today's technologically savvy students expect the use of technology in their college courses. A study of the use of technology at the South Texas Community College Starr County campus found that students believed having access to technology in a class helps them learn regardless of whether or not they actually even used the technology.[2] Higher education administrators, with increasingly tight budgets, are looking to online courses to reach more students with fewer facility dollars being spent. Online courses are accessible by mobile devices, allowing students to access course materials anywhere, anytime. In spite of the pressure to use online educational tools, many faculty members are hesitant to adopt online courses. Students' perceptions of the courses that are offered online are not always favorable either. Clearly, education has lagged in adopting new technology that would allow greater mobile access to educational materials.

Faculty Perceptions

In a literature review of studies of faculty barriers and motivation for adopting online courses, the author found both intrinsic and extrinsic reasons that faculty were hesitant to participate in distance education courses.[3] Intrinsic fears of online education included resistance to change, technology intimidation, fear of traditional courses being replaced by online education, and worry over intellectual property issues. Extrinsic reasons included feeling that course quality would be sacrificed, inappropriate for older more traditional students, decreased student interactivity, copyrights issues, faculty workload, unreliable technology, lack of training, and lack of recognition for online teaching. Extrinsic concerns outweighed intrinsic ones, and the study concluded that if educational institutions addressed some of the extrinsic problems, then the intrinsic concerns would be lessened.

Motivations for adopting online courses were also explored in the literature review. Intrinsic motivators, which were found to be stronger than extrinsic motivators in this case, included a personal desire to use technology in teaching, job satisfaction and self-gratification, and optimal working conditions since online courses can be taught anytime, anywhere. Extrinsic motivators included support and recognition especially toward tenure, peer pressure, getting students involved in technology, enhancing course quality, and providing better access for students at a distance. With support from institutional administration, training, and peer modeling, the study concluded that faculty members would be more inclined to participate in online education.

Another drawback to a faculty member's participation in online education may be his or her particular pedagogical beliefs. Online courses require the instructor to be more of a facilitator rather than a traditional lecturer. Many faculty members are uncomfortable with this new role. Training in facilitation techniques may help to create a greater comfort level with this type of instruction. Some faculty members may be more successful at online education than others who are less comfortable with the facilitator role, which gives more power and responsibility for learning to the students. Unfortunately, the adoption of content management systems in academic institutions often happens without including the library and librarian faculty members in the design and implementation of these systems. In a 2009 article on the online classroom and librarians, the authors encouraged academic librarians to get involved and integrated into their institutional course management systems (CMS), including lobbying for a library link embedded in the course management system.[4]

Student Perceptions

While students are typically used to having technology in their everyday lives, researchers have found that they may not be particularly motivated to participate in online courses. In a 2010 study on students' experiences in e-learning, researchers found that students preferred face-to-face learning for establishing social relationships with other students and for cooperative learning experiences.[5] They felt that face-to-face learning was better for acquiring knowledge, skill building, and application of knowledge. Students experienced with online learning felt that it was better for structuring and organizing knowledge. They also appreciated the cost savings from not having to travel to campus. Students also appreciated the fast feedback from instructors that online courses can provide. Students felt that online learning was more flexible and provided more opportunities for practice exercises. They also appreciated the increased opportunity to self-monitor their progress and for self-paced learning. Mobile education also allows students in remote areas who do not have access to a physical classroom to access education online. Library information literacy, which is typically an add-on to academic coursework, lends itself well to online learning formats since library modules can be created separately and added on to course websites.

Librarian Perceptions

In a 2008 article on promoting critical thinking skills for lifelong learning in online information literacy instruction, the author strongly advocated for using the case-based, constructivist learning model for online library instruction.[6] The goal of information literacy instruction is to produce students who can find, evaluate, manage, and effectively use information when needed. Critical thinking skills are necessary for being able to navigate the increasingly complex world of online information resources integral to information literacy today. Constructivist learning models help develop critical thinking skills in students and are therefore highly desirable for teaching information literacy. In addition, since so many of the library's resources are now online, it makes sense to teach information literacy online as well. Helping students become independent and critical learners is central to the mission of higher education, and librarian-faculty partnerships are essential to develop lifelong research and education skills in students. The convenience of student access to library learning modules at the point of need, allowing them to work at their own pace, combined with the ability to reach distance learners makes online learning highly desirable for library information literacy instruction.

TYPES OF MOBILE LEARNING

Distance learning can be conducted in two ways: asynchronous and synchronous. Asynchronous online learning is the most common online course method. Using this method, students follow a prescribed course of study at their own pace and on their own time schedule. This type of distance learning is often accomplished through the use of a CMS such as Blackboard, Canvas, and Moodle or through the creation of static web pages. Most modern course management systems are accessible from mobile devices, as are teacher-created websites. Blackboard has the Mobile Learn app to connect to online courses created in the Blackboard CMS. It is available for iOS and Android. Moodle and Canvas also have apps available for iOS and Android (figure 8.1).

The teacher or facilitator of an online course creates lessons or modules, and the student completes the required coursework and assessments. There are opportunities to interact with faculty and other students through email or discussion board postings, but these interactions are limited since there is no immediate feedback. Synchronous online learning occurs when technology such as web-conferencing software and chat rooms are used so that students can interact and collaborate with other students and faculty members in real time. This type of online learning allows more immediate feedback and interaction between students and faculty and is often accomplished by using web conferencing tools. Synchronous learning should be used when interactions between course members are crucial for a particular learning objective.

Figure 8.1. Canvas by Instructure mobile app

A study by the University of Southern Mississippi showed that students rated synchronous online activities significantly higher than asynchronous online learning.[7] Students felt that synchronous online learning was closer to traditional face-to-face learning in that it encouraged contact between students and with their teachers. Students also felt that it encouraged more active learning and offered more diverse ways of learning than asynchronous learning methods. In a study on online cooperative learning, the authors found that student achievement was shown to increase with the introduction of online structured group collaboration activities.[8] Synchronous technologies can help create more of a community among students in a course than the traditional asynchronous method of online learning, and can be accomplished through the use of chat and group discussion online. Mobile tools such as CMS apps, Skype, and even social networking apps such as Facebook can be used to facilitate information sharing and online learning accessible by mobile technology. Cloud applications such as Google Docs and Dropbox facilitate remote sharing of documents and group work. Web conferencing, which is built into the Canvas CMS, allows synchronous learning right from within the online course learning environment. All of these tools are readily available to librarians, especially in an academic setting.

The increase in online courses as well as in hybrid courses, where part of a course is hosted online and part is face-to-face, allows greater access to academic coursework from mobile devices than ever before. In a 2013 survey at the University of Central Florida, researchers found that students are increasingly using smartphones and tablets for academic pursuits.[9] Students used mobile devices for learning in multiple ways, including accessing content management systems, using online learning programs, looking up reference information, reading e-books, and producing documents and presentations. Apps used most frequently included Khan Academy, iTunes U, CourseSmart, Inkling, Dictionary, Wikipanion, WolframAlpha, Evernote, Dropbox, Pages, Keynote, and Notes.

Abilene Christian University launched an institution-wide mobile learning initiative in 2008. Their goal was to "imagine a world where classes become untethered from the stony isolation of four walls, where information is accessible in new contexts and situations, where learning becomes truly mobile, permeating our students' lives."[10] ACU implemented the program by giving iPhones to students, faculty, and staff and providing training and learning opportunities throughout the academic year. The program was entirely voluntary and the iPhones given to students were considered their personal devices; usage for academic pursuits alone was not mandated. The students used the devices for social networking, communication, and entertainment and the device became an integral part of the students' lives. The first year of the program proved so successful that the institution continues to provide the necessary resources to continue the program, and even expanded the program to the use of iPads.

Besides using mobile technology to teach traditional subjects, faculty and students at Abilene Christian University have participated in many innovative programs such as working with area elementary school students using mobile technology, using mobile technology when working in the field for recording and measuring scientific information, creating mobile apps to facilitate teaching and learning, and creating art using mobile apps such as

Sketchbook and Penultimate. In addition, many courses have moved to using exclusively electronic textbooks and submitting online assignments, making the education experience almost totally paperless. In their studies of the use of the iPad with online education, researchers found that the majority of students felt that iPad usage in a course facilitated more active learning. Students also indicated that using an iPad increased interaction with peers outside the course, which also enhanced their learning experience.

DESIGN OF ONLINE COURSES FOR MOBILE ACCESS

In addition to deciding which technology solutions to use, there are many considerations when designing an online course. In the 2010 study of student preferences in e-learning mentioned previously, the authors identify several instructional components that should be included in a high-quality online course offering.[11] First, courses should be designed with explicit learning outcomes. Learning outcomes can be conceptual and methodical, but also should include social and personal outcomes such as teamwork and self-directed learning. Second, studies have shown that a logical structure and outline to an online course will positively affect student satisfaction with the course. The technology used will also positively or negatively affect the structure and this should be taken into consideration when choosing a learning management or delivery system. For example, the Canvas CMS has a Modules component that can be used to organize the class by weeks or sessions so that students can clearly see what is assigned for a particular time period. Third, the presence of a skilled instructor is a key course component to provide structure, give feedback, stimulate motivation, provide assistance, and help students reflect on content. Dropout rates are typically found to be high with online courses, and faculty-student interaction can positively affect motivation, engagement, and satisfaction, leading to lower dropout rates. Fourth, interaction with fellow students is also important for mutual support, engagement, and satisfaction, so activities such as chat and discussions are crucial. Finally, online learning allows for individualized student learning processes. Students should receive ample opportunities to practice and apply what they learn and develop skills for self-regulation of learning. Also, course management systems allow for online quizzes, which give feedback to students and provide opportunities to practice learning.

One educational method that works well for online and therefore mobile instruction is problem-based learning (PBL), also known as case-based learning. PBL can be defined as a "learner-centered instructional approach that empowers learners to think critically; analyze and solve complex, real life problems; find, evaluate and use appropriate learning resources; work collaboratively; demonstrate effective communication skills; and become lifelong learners."[12] This definition incorporates the main goals for library information literacy training. It is also particularly valid for graduate medical education where students are developing their thinking, evaluation, and diagnostic skills. Another study endorses problem-based learning for developing valuable work skills such as problem solving, adaptability, and analytical skills.[13] PBL can help develop such transferrable skills as critical thinking, teamwork, and self-directed learning. Many studies have also shown that PBL works best when the case is related to actual real problems that students may encounter in current or future work, also known as "authentic learning assignments."

Problem-based learning is usually practiced in traditional face-to-face classrooms, but can be modified for a more self-directed online approach. There are several steps that should be taken when designing a PBL learning environment. The first steps guide the student through activating their prior knowledge and elaborating on that knowledge. Next, students analyze and classify their knowledge to discover the gaps in their learning, followed by self-directed searching for knowledge to fill the gaps. Finally, students inform each other and evaluate their own learning. In another study on the use of PBL in online courses, the authors use Bloom's taxonomy to scaffold learning.[14] Bloom's taxonomy encourages higher-order thinking such as application of knowledge, synthesis, and evaluation. Problem-based learning can be broken into three basic steps: "defining the issues," "gathering information," and "solving the problem." Problem-based or case-based learning is usually best done in small groups of three to five students, so larger classes should be divided up into groups for optimal learning.

Introducing technology into problem-based learning significantly changes the role of the instructor. In a study of computer-supported collaborative learning, the authors found that introducing technology in the form of software and an interactive whiteboard allowed the instructor to have more time for "engaging students in challenging situations by increasing the complexity of the problem."[15] The technology served as "scaffolding" or support for learning so that the instructor's role involved less explaining and prompting, which was handled by the software, and more guidance in higher-learning processes. An online course management system can also serve as scaffolding for learning in this way. PBL follows the constructivist model of guiding students through their own learning. In a paper exploring the role of the lecturer in problem-based learning in health education, the author found that successful PBL instructors needed to be able to assume the role of being a guide in order to let students construct their own knowledge, and avoid imposing their own point of view on the students.[16] A successful PBL instructor knows when to intervene to facilitate discussion, but only when the student process loses direction. This is a learned skill and not all instructors are comfortable with this model.

According to a 2008 article on educational technology and librarianship, facilitating learning with technology is a core function of librarians.[17] Information literacy instruction, online education, and the design of instructional materials are all skills needed by librarians today, especially in an academic setting. Although there is little formal educational training in library schools, graduates are expected to have familiarity with technologies used to facilitate and improve the performance of students. In reality, many new librarians are unequipped for the teaching role. In a 2013 article on the librarian teaching role, the author states that "librarians must be positioned as key educators in the teaching and learning environments of the future."[18] This will require librarians to learn new skills and concepts, and to become educational partners with faculty to integrate information literacy skills throughout the curriculum. Some of the skills needed include knowledge of learning theories, instructional design, the ability to develop instructional materials, and the ability to assess learning. Relevant and timely professional development is crucial to this goal.

When designing educational programs for libraries using mobile technology, we also need to know our user community and their mobile use patterns beyond their instructional needs. We know that, especially in academic settings, more and more of our users do not come into the library for in-person instruction, so we are increasingly providing online tutorials from our websites. Web analytics can be used to determine the most commonly used devices when accessing the library website. Surveys of users can also be used to uncover mobile use patterns and instructional needs. Observations of mobile usage in the library and classrooms can also provide helpful information and guidance in designing successful library educational modules for mobile learning. We know from educational research that students prefer synchronous learning, and typical library tutorials are asynchronous. Using a CMS with built-in tools such as chat, discussion boards, and online quizzes may be more successful. Problem-based learning should also be considered as a method for successfully providing library instruction.

At Boston University, librarians and faculty members from the Department of Family Medicine worked together to create a web-based curriculum to teach evidence-based medicine (EBM) to third-year medical students.[19] In many medical school programs, clerkships are held in community clinics and other off-site placements, so it is difficult to standardize training for all students. An online training program solves this problem. The online course included interactive elements such as reflective learning and collaborative and peer-driven activities, and was taught using the Blackboard CMS. Application of real patient encounters to course concepts was also incorporated into the curriculum, along with an asynchronous discussion board. In a study comparing medical students who completed the online course and those who completed tradition training, the online group scored higher, indicating greater improvement of searching skills after the training. Post-training survey results also indicated that students who completed the online course exhibited more confidence in their ability to perform an evidence-based literature search.

INTEGRATING MOBILE TECHNOLOGY IN EDUCATION

There are many ways to begin to integrate mobile technology into educational programs. Blogging software such as WordPress now comes with built-in mobile-friendly templates. Students can access class blogs from their mobile devices to make postings during a field trip or sports event, or even while riding the bus. Podcasts of lectures can be loaded on iPods for checkout in the library, or made available on a website for download to personal mobile devices. Apps have been created for gathering scientific data in the field, enabling students to use mobile devices for data collection. Mobile devices have also been used in the field for gathering feedback from surveys or impromptu polls of people in public places. Educational games and scavenger hunts can be played on students' personal devices. The built-in GPS features of mobile devices can be used for providing location-based information in educational programming, such as tours of cultural sites. This type of educational programming is approaching augmented reality, which is the idea that information can be provided as you move through spaces or areas based on your location. There are now many apps that provide travel, dining, shopping, and cultural landmark information on your mobile device using the location-aware feature. This information can be provided on your phone or tablet as well as through new devices being developed such as Google Glass, available for purchase since summer 2014.

In a 2012 article written by nurse educators, the authors describe an innovative program they developed to integrate the use of cell phones into their classroom activities in order to make classes more interactive and engaging.[20] The authors found that all students in their classes owned cell phones, although not all of them owned a smartphone. Some activities were held in groups in order to accommodate the students who did not have a smartphone. Three types of student-centered cell phone learning activities were used. The first involved texting a friend with a quick question to do an informal and anonymous survey on sensitive topics such as eating right and religious, cultural, or sexuality questions. Another activity encouraged students to use smartphones to enrich learning topics by looking up information on the web. Topics to be searched can be provided or a scenario proposed, and students use online resources to look for solutions. Similarly, students can use their smartphones to look up patient education materials related to a particular topic, or access online care or practice guidelines. This type of activity could be greatly enhanced by using resources selected by librarians to search for the information. The third type of activity involved using online survey software such as Poll Everywhere or SMS Poll with cell phones as a personal response system. Multiple choice questions can be asked, as well as short-answer questions relating to evaluating class learning or asking for further information. The authors instituted class rules on appropriate cell phone usage and the students responded favorably to the activities, reporting that the use of mobile technology kept them active and involved in course learning.

USING MOBILE TECHNOLOGY IN LIBRARY EDUCATION PROGRAMS

As a result of increasing mobile technology use by students, library instructors are using mobile technology in the classroom more often, as well as making sure our online materials and tutorials are mobile-friendly and our electronic resources are mobile-accessible. Librarians can use mobile technology to teach directly from wireless tablet device with cables to connect to a projector or screen. When in non-library classrooms, research labs, clinics, or other areas where librarians can be embedded, a mobile device can be indispensable to library work. Librarians can use mobile devices to take notes, update teaching materials, and access resources to demonstrate to students, as well as look up information at the point of need or display a prepared presentation for teaching purposes.

Some ideas for teaching and learning in libraries with mobile devices include tablets or iPods available for checkout in the library, pre-loaded with apps for accessing library databases, audio lectures, or music files; e-readers loaded with e-books and PDF materials available for checkout; interactive orientations and tours of the library that employ QR codes readable by phones or tablets that link to a website describing resources available at that location; and use of software such as Poll Everywhere that allows students to respond to a question using their personal mobile devices. Online polling is a good way to get the audience involved and engaged in large orientation sessions. Polling can also be used for formative and summative assessment during library teaching sessions to determine how students are doing during the session, and again at the end of the session to determine if students have met the learning goals.

Library instruction can be made much more interactive and engaging for little or no cost using mobile technology that most students have at hand. Case-based learning can be used with mobile devices by presenting a scenario and having students find the answer using library apps for database searching. Blogs and social media apps such as Facebook or Twitter can be used to augment learning and facilitate discussions around class topics, although care needs to be taken to ensure that public postings are not offensive or hurtful to students; posting should be mediated by a facilitator. At the University of Washington Health Sciences Library, we are developing a serious game using patient case scenarios to teach evidence-based medicine. A beta game was produced in 2013 and previewed by a small audience at a regional conference. While some of the participants brought laptops to the conference, others brought tablet computers. We quickly realized that some elements of the game did not work well with touch technology. We are now redesigning the game so that it is optimized for play on mobile devices.

Be aware that there can be some drawbacks to using mobile technology in classroom settings. Students can engage in off-task behavior by looking at email or social networking sites during a lesson. Many K–12 settings prohibit the use of personal mobile devices in classrooms. But if designed correctly, students can use mobile technology responsibly in the classroom and be held accountable by the teacher to be on task and complete their assignments. In a K–12 setting, school-owned devices may be used to avoid personal distractions. There are programs such as Fakebook for use by K–12 educators to safely use social networking without the security risks. Another concern is wi-fi or cellular connections that are weak or not available in certain library areas or classrooms. Infrastructure upgrades may be necessary before implementing a mobile learning environment in your library or classrooms. Educational grants or awards from local or government agencies may be available to help with this problem. Mobile communication is here to stay, and growing. What better way to reach and engage students than to use the technology that they are already using in their daily lives?

When using mobile technology in education, there is no need to completely reinvent the wheel. Existing online content can be adapted for use with mobile devices. Some important considerations for adapting existing content for mobile device use include:

- Allowing for smaller screen size
- Breaking up educational materials into short topics and making lessons modular
- Allowing for touch-based navigation when designing online materials
- Using problem-based or case-based learning
- Testing tutorials and online materials with many devices to ensure compatibility with all devices and browsers

Online tutorials can be broken into short modules for mobile-friendly access or hosted on a WordPress or LibGuides site with a built-in mobile template, or a course management system with a mobile app, such as Canvas. Beginning with an assessment of your library's user needs, and focusing on mobile educational programs that will best serve your unique library users, will ensure a successful implementation of a mobile educational program in your library.

NOTES

1. New Media Consortium. “Sparking Innovation, Learning, and Creativity.” Horizon Report: Higher Ed Edition, NMC, 2012. http://www.nmc.org/publications/horizon-report-2012-higher-ed-edition (accessed May 30, 2014).
2. South Texas Community College. *Student Perceptions of the Use and Educational Value of Technology at the STCC Starr County Campus: Implications for Technology Planning*. Office of Institutional Research and Effectiveness, STCC, 2002.
3. Maguire, Loréal L. “Literature Review: Faculty Participation in Online Distance Education: Barriers and Motivators.” *Online Journal of Distance Learning Administration* 8, no. 1 (2005). http://www.westga.edu/~distance/ojdla/spring81/maguire81.htm (accessed May 30, 2014.
4. York, Amy C., and Jason M. Vance. “Taking Library Instruction into the Online Classroom: Best Practices for Embedded Librarians.” *Journal of Library Administration* 49, nos. 1–2 (2009): 197–209.
5. Paechter, Manuela, and Brigitte Maier. “Online or Face-to-face? Students’ Experiences and Preferences in e-learning.” *Internet and Higher Education* 13, no. 4 (2010): 292–97.
6. Allen, Maryellen. “Promoting Critical Thinking Skills in Online Information Literacy Instruction Using a Constructivist Approach.” *College and Undergraduate Libraries* 15, no. 1 (2008): 21–38.
7. Ward, Michael E., Gary Peters, and Kyna Shelley. “Student and Faculty Perceptions of the Quality of Online Learning Experiences.” *International Review of Research in Open and Distance Learning* 11, no. 3 (2010): 57–77.
8. Nam, Chang Woo, and Ronald D. Zellner. “The Relative Effects of Positive Interdependence and Group Processing on Student Achievement and Attitude in Online Cooperative Learning.” *Computers and Education* 56, no. 3 (2011): 680–88.
9. Chen, Baiyun, and Aimee Denoyelles. “Exploring Students’ Mobile Learning Practices in Higher Education.” Educause Review, October 7, 2013. http://www.educause.edu/ero/article/exploring-students-mobile-learning-practices-higher-education (accessed May 30, 2014).
10. “ACU Mobile Learning Report 2008–2009.” Abilene Christian University, 2008. http://www.acu.edu/technology/mobilelearning/documents/acu-mobile-learning-report-2008-09.pdf (accessed May 30, 2014).
11. Paechter and Maier, “Online or Face-to-face?”
12. Baturay, Meltem Huri, and Omer Faruk Bay. “The Effects of Problem-based Learning on the Classroom Community Perceptions and Achievement of Web-Based Education Students.” *Computers and Education* 55, no. 1 (2010): 43–52.
13. Kivela, Jakša, and Ruth Jeanine Kivela. “Student Perceptions of an Embedded Problem-based Learning Instructional Approach in a Hospitality Undergraduate Programme.” *International Journal of Hospitality Management* 24, no. 3 (2005): 437–64.
14. Cheaney, James D., and Thomas Ingebritsen. “Problem-based Learning in an Online Course: A Case Study.” *International Review of Research in Open and Distance Learning* 6, no. 3 (2005). http://www.irrodl.org/index.php/irrodl/article/view/267 (accessed May 30, 2014).
15. Lu, Jingyan, Susanne P. Lajoie, and Jeffrey Wiseman. “Scaffolding Problem-based Learning with CSCL tools.” *International Journal of Computer-Supported Collaborative Learning* 5, no. 3 (2010): 283–98.
16. Haith-Cooper, Melanie. “Problem-based Learning within Health Professional Education: What Is the Role of the Lecturer? A Review of the Literature.” *Nurse Education Today* 20, no. 4 (2000): 267–72.

17. Johnson, Wendell G. "Educational Technology and College Librarianship." *College and Undergraduate Libraries* 15, no. 4 (2008): 463–75.

18. Peacock, Judith. "Teaching Skills for Teaching Librarians: Postcards from the Edge of the Educational Paradigm." *Australian Academic & Research Libraries* 32, no. 1 (2001): 26–42.

19. Schilling, Katherine, John Wiecha, Deepika Polineni, and Souad Khalil. "An Interactive Web-based Curriculum on Evidence-based Medicine: Design and Effectiveness." *Family Medicine* 38, no. 2 (2006): 126.

20. Robb, Meigan, and Teresa Shellenbarger. "Using Technology to Promote Mobile Learning: Engaging Students with Cell Phones in the Classroom." *Nurse Educator* 37, no. 6 (2012): 258–61.

RESOURCES

A Guidebook for Social Media in the Classroom: http://www.edutopia.org/blog/guidebook-social-media-in-classroom-vicki-davis

Library Uses QR Codes for Customized Tour: http://www.emory.edu/EMORY_REPORT/stories/2011/06/campus_woodruff_library_coca_cola_robert_woodruff_qr_code.html

Peters, Thomas A., and Lori Bell. *The Handheld Library: Mobile Technology and the Librarian*. Santa Barbara, CA: ABC-CLIO, 2013.

Poll Everywhere: http://www.polleverywhere.com/

Top 10 Augmented Reality Travel Apps: http://travel.cnn.com/explorations/life/top-10-augmented-reality-travel-apps-569570

Chapter Nine

Facilitating Outreach and Communications with Mobile Devices

Library outreach services and programs may benefit the most from using mobile technology. Outreach services, since they are typically done outside of the physical library, are great places to start with mobile technology initiatives. Libraries have a long tradition of doing outreach to community and special populations, but outreach with mobile devices has not been widely adopted by many libraries. Similarly, mobile devices have much potential as a communication device for librarians who do outreach or are embedded in academic departments or research labs. Yet initiatives for getting mobile devices into the hands of librarians to facilitate outreach have not been widely accepted.

Library outreach is defined by Wikipedia as "an activity of providing services to populations who might not otherwise have access to those services."[1] The key to outreach is meeting the informational needs of people where they are. This logically means that outreach must be done outside of physical library spaces. Mobile technology is the perfect tool for meeting the outreach needs of any type of library. Public libraries serve all members of the community including the elderly, disabled, incarcerated, and homeless populations. Academic libraries serve students of all kinds, whether they live on campus, commute to campus, or attend courses completely online. Special libraries often serve even more diverse and far-flung populations such as those associated with museums and historical societies with collections of interest to populations around the world. The term *outreach* is also sometimes used to mean marketing efforts, which is valid for most libraries because outreach serves to carry the message of library services and programs to a much wider audience than just those who visit the physical library building.

The definition of outreach can be different for every library, just as the library user population, collections, and services are all different. In a 2005 article about health sciences library outreach, the authors recommended that every library create a clear definition of what outreach means to them, considering specific program goals and objectives in order to select the best outreach activities on which to focus.[2] Librarians should clearly understand the objectives of outreach for a particular community in order to ensure the best use of library funds and successful outreach initiatives. Funding for library outreach initiatives is crucial and often limited, so priorities will need to be created. Other barriers to success can be failure to coordinate with other community members and assess needs before launching an outreach program. It is crucial to coordinate with community members and seek input and collaboration from all stakeholders in order to be successful with outreach initiatives.

As with all library programs, assessment of user needs is crucial before beginning any new outreach or communication initiative. In an article describing outreach programs to underserved students at Oakland University, the authors emphasize the importance of taking the time to accurately assess user needs in order to create effective outreach programs.[3] Building relationships with community members is the most accurate way to assess user needs. Library assessment programs should go beyond surveys and focus groups to direct interactions with users in their environment in order to truly assess their information needs. That means going beyond the four walls of the library to where users are working and studying. Building partnerships with other service departments in organizations is another way to discover outreach needs. Through partnerships we not only get new ideas, but also share resources and establish the library as a collaborative entity, as well as gain more visibility for the library and its programs.

LIBRARY OUTREACH PROGRAMS USING MOBILE TECHNOLOGY

Background

In many ways mobile devices are no different from other portable technologies such as laptops because they are just tools to support librarians' roles in teaching, learning, research, and outreach to the greater community. Smaller, more portable but still powerful technology expands our ability to share and present information beyond the four walls of the library in much more flexible and innovative ways. Using mobile devices in library outreach and communication programs gives librarians optimum mobility, plus the ability to engage with academic and community members in new and much more interactive ways. Using mobile technology allows us to present information at the point of need, when the information is actually being used, not just in an artificial, academic environment. When librarians are embedded in users' environments, they can become much more familiar with and better understand users' information-gathering processes. This understanding also leads to better anticipation of future information needs for the library community.

In an article describing outreach programs in vocational educational settings, the authors emphasized that in our exponentially expanding informational world, users' expectations for immediate service are changing.[4] These expectations are making the location of library services a crucial consideration for keeping libraries and librarians current and relevant to users. The model of the embedded librarian who is incorporated into a classroom, department, or research environment is not a new idea. In the modern information environment, it is crucial that librarians break out of the passive role of waiting for users to come to the library for services and reach out to the institutions and communities of which they are a part. Librarians must proactively seek integration into teaching and learning, as well as into community activities, in order to provide value and return on investment or risk losing funding in competitive budgetary times. Embedding librarians in the communities they serve also fosters trust and mutual understanding with community partners, faculty, and administrators.

Outreach programs can create crucial partnerships for libraries with important community stakeholders. Outreach also makes the library more visible to the community of which it is a part and therefore increases its value to the community. Creating relationships with stakeholders through outreach drives innovation because ideas for serving the community arise from working with others. Outreach programs address the "big picture" needs of a community or organization. Only through working with others can we truly understand their information needs. Aligning library services and programs with community values and mission will help ensure successful collaborations. Partnering with community members can also save the library money by sharing costs for programs. In these days of diminishing budgets, this consideration cannot be overlooked.

As reported in the library literature, the types of outreach activities conducted by libraries vary widely according to types of libraries and the types of communities they serve. Some populations that have benefitted from library outreach include medical staff in hospitals and clinics, research and laboratory personnel, underserved student populations, homebound and disabled community members, remote populations, the homeless, and other special needs populations such as the elderly and incarcerated individuals. We will examine each of these populations in turn and explore example outreach programs using mobile technology, which can be effective for providing outreach to that specific population.

Outreach to the Homeless and Underinsured Individuals

Librarians at the University of Washington Health Sciences Library were recently invited to participate in Washington Care Clinic, a community health fair sponsored by the National Association of Free and Charitable Clinics.[5] This large-scale, all-day event provided free medical care to hundreds of underinsured individuals. Many physicians, nurses, and dentists volunteered to serve at the clinic. Library volunteers provided health care information and referral services to individuals who needed more information after visiting the health care workers. Library volunteers staffed a booth with a laptop computer, printer, pamphlets about Medline Plus, and two iPads. Use of the iPads allowed volunteers to move around the exhibit area and interact with people needing disease and treatment information as well as those needing information on finding follow-up services. So many people were looking for dental services that the dental clinic filled up early, so there was a pressing need to find affordable or free dental care services for those who did not get an appointment that day. The local 211 information and referral services website was used to look up affordable or free health care and dental services. Some communities in California, Florida, and South Carolina now have apps available for 211 information searches. We also referred patients looking for disease information to MedlinePlus, which has a mobile website. The iPads proved invaluable to the success of this large-scale community outreach event.

Outreach with Underserved Student Populations

Serving the needs of the diverse student body on today's college campuses can be challenging for academic librarians. Diverse student populations, distance learning, and vocational education all present challenges for librarians providing outreach services. The many types of programs that can be served by library outreach programs include programs for new and transfer students, multicultural and travel abroad programs, disabled student services, honors and advanced scholar programs, residence hall outreach programs, service to teaching assistants (TAs) and research assistants (RAs), community practicum programs, and athletic tutoring, as well as distance and vocational education. Through assessment and input from institutional administration and student programs, academic libraries will be able to identify which student populations are in greatest need of library outreach services. Outreach services can also be used to market library programs and expand services to students who previously did not use library services. In a 2008 article about outreach programs to students, the authors identified and piloted emerging technologies and outreach strategies with the help of campus partnerships.[6] They chose library exhibits using multimedia kiosks, book talks, and outreach to TAs as important initiatives for their institution. All of these programs would be greatly enhanced by the use of mobile technology. For example, tablet devices can be loaded with multimedia presentations for complementing a library exhibit. Tablets can also be loaded with e-books and research papers supporting the exhibit topic. Access to blogs for book talks can be bookmarked on iPads for students to check out, and librarians can use tablet devices when performing outreach to busy TAs and RAs by visiting them in their work spaces to demonstrate helpful databases for their subject area and search techniques for use in their research and teaching.

Outreach Programs with Medical Residents and Clinical Faculty

Serving the needs of medical staff at academic health sciences libraries is also challenging. Nurses, doctors, and other allied health care professionals are very busy and rarely have time to visit the physical library, although online access to evidence-based practice resources is crucial to their effective work practices. Librarians who conduct outreach in medical facilities find mobile technology to be extremely helpful in making their work more efficient. Clinical librarians at the University of Washington Health Sciences Library bring iPads to case conferences, where residents and attending physicians discuss interesting cases they have encountered during rounds.[7] Using apps such as Evernote, librarians take notes on the cases presented and can look up information to report to the group if needed. Bluetooth keyboard cases can be purchased for the iPads to make taking notes easier. Both Zagg and Kensington have several popular and reliable models to choose from. The Evernote app is great for taking notes and is a cloud app, so notes can be accessed later from any computer or browser. Librarians had earlier posted clinical questions that arose from clinical sessions on a password-protected blog site with links to evidence-based articles addressing these questions. The blog then becomes a database of clinical cases and journal articles that can be accessed by everyone in the group or program. This same method can also be used successfully with clinical departments, such as nursing or psychiatry. Librarians attend case conferences or departmental meetings and take notes, then later post literature addressing the questions that arise during the conferences on a secure blog that is set up for communications with the department.

In addition to participating in resident conferences and departmental case discussions, demonstrations on the use of tablets with medical apps are extremely popular with clinicians. Librarians can review and highlight useful clinical apps on the library website, and then demonstrate their use in clinical cases as an outreach project in a medical center or clinic. Some categories of apps that can be demonstrated include apps for accessing medical databases such as Dynamed and Micromedex, apps for clinical calculations such as

Calculate by QxMD and AnticoagEvaluator, apps for reading journal articles such as Browzine or Read by QxMD, and specialized apps such as VisualDX, Radiology 2.0, and ARUP Consult. To display the information to the group, a tablet computer can be connected to a projector with an adapter cable or directly to a large screen monitor. Residents and attending faculty at the University of Washington Medical Center have been very enthusiastic about learning to use tablets and apps. Live app demonstrations go much smoother when the apps are arranged in folders by function, such as drugs, diagnosis, and utility apps. A slide show of sample cases can be used to demonstrate the use of the apps. Saving the PowerPoint to PDF format and then downloading it to iBooks is helpful for running the slide show directly from the iPad. Switching back and forth from the PDFs of cases in iBooks to apps loaded on the iPad is easier than using the browser. This approach seems to work well and be fairly stable for demonstrations. There are programs available that mirror the iPad screen on a computer that is connected to a projector using the built-in AirPlay feature, but it may be more reliable to work directly from the iPad. Another approach is to use software such as Reflector to capture the iPad screen and save screenshots to insert into a static slide show presentation.

Outreach to the Elderly

If your library community has an active elderly population, you might consider an outreach program to seniors using mobile technology. According to ZDNet.com, the simple interface and small form factor of the iPad mini makes it the perfect fit for elderly users.[8] Offering iPads in the library especially for seniors to check out, to read articles or use apps of interest to them, could become a popular community service. According to the website MyAgeingParent, studies have shown that seniors are able to read faster using an iPad or Kindle device than from regular books.[9] This may be because it is so easy to adjust the size of the font on a mobile device. Using an iPad might also assist memory and encourage social skills. Seniors can use a tablet device to shop online, get information, and connect with distant family and friends. Besides reading materials, tablets can be loaded with games, music, and videos applicable to any taste. Some of the top free apps recommended for seniors include:

- iBooks (iOS): https://www.apple.com/ibooks/
- Lumosity (iOS and Android): https://itunes.apple.com/us/app/lumosity-mobile/id577232024?mt=8 and https://play.google.com/store/apps/details?id=com.lumoslabs.lumosity&hl=en

- Netflix (iOS and Android): https://itunes.apple.com/us/app/netflix/id363590051?mt=8 and https://play.google.com/store/apps/details?id=com.netflix.mediaclient&hl=en
- NPR (iOS and Android): http://www.npr.org/services/mobile/iphone.php and http://www.npr.org/services/mobile/android.php
- Skype (iOS, Android, Windows Phone, BlackBerry): http://www.skype.com/en/download-skype/skype-for-iphone/, http://www.skype.com/en/download-skype/skype-for-android/, and http://www.skype.com/en/download-skype/skype-for-mobile/
- The Weather Channel (iOS, Android, Windows Phone, BlackBerry, Kindle Fire): http://www.weather.com/services/mobilesplash.html
- Words with Friends (iOS and Android): https://itunes.apple.com/us/app/words-with-friends/id321916506?mt=8 and https://play.google.com/store/apps/details?id=com.zynga.words&hl=en

Outreach to Incarcerated Individuals

Incarcerated individuals have some of the lowest health literacy levels in the United States. There is a pressing need to address these issues through mental and physical health education in order to prepare these individuals for a return to society and to better manage their health and reduce the burden on the communities that support them. In a recent outreach project, a consumer health librarian collaborated with community members to provide health information training classes to inmates in the Gallatin County Detention Center.[10] Inmates are not allowed access to the Internet during their incarceration, so all training materials had to be created and saved to computers offline. The training program was ambitious—trying to reach 70 inmates in 5 months with 10 classroom sessions each—and the detention center only had 7 computers available to use. Due to disqualifications and interruptions in inmate schedules, only 39 individuals completed the training, but the ones who did complete the program remarked on its helpfulness and indicated they had learned better health care management. This kind of outreach program could easily be enhanced with the use of tablet computers, by downloading all of the offline materials as well as nutrition calculators and other health tools that do not require an Internet connection to run. With the use of tablet computers, it might be possible to reach many more inmates in a shorter period of time.

Embedded Librarians

Embedded librarianship has been discussed in the library literature for many years and can mean different things in different settings. In a 2010 article introducing the topic, the authors defined embedded librarian programs as those locating “the librarians involved in the spaces of their users and colleagues, either physically or through technology, in order to become a part of their users’ culture.”[11] The authors pointed out that embedded librarians understand and work closely within their user populations. Close proximity to their users’ work places is key to understanding their needs and being a part of the community. Collaborations with administrators and managers within the user population are also very important to maintaining a prominent role for librarians within academic and other institutions. The authors concluded that many libraries are embedding librarians in the community, using technology to bring services to larger audiences, and predicted that embedded librarianship will grow as technology advances. With these advances, embedded librarianship is greatly enhanced by freeing up librarians to fully integrate their services into the wider community.

At Johnson and Wales University, North Miami Campus, experiential learning is a focus.[12] There are no predetermined library requirements for courses, so librarians must reach out and establish relationships with departments in order to provide services to students. Embedded librarians at Johnson and Wales are immersed within the community and culture of the academic departments they serve. Librarians are embedded in the community in several ways: as satellite librarians located in strategic campus areas at key times during terms, at career conferences where they create informational web pages for students, during the first-year experience program, and in classrooms in specific departments such as the culinary arts program. Using this model, librarians are embedded in the entire campus community, serving students throughout their academic experience. Library programs become more proactive than reactive through outreach to users at the point of need. Mobile technology is essential to this model of librarianship. This level of embedding can be done with the use of a laptop, but smaller, lightweight tablets with a longer battery life are much more flexible for working directly with students in campus common areas and classrooms.

At the National Cancer Institute–Frederick Scientific Library, librarians found that fewer people were leaving their research buildings to come to the physical library.[13] In order to be proactive, they decided to start offering in-person “Laptop Librarian” services. Librarians were stationed in areas highly visible to researchers within their labs and answered questions from their laptop station, but found that it was sometimes necessary to go to the researcher’s desk to answer a specific question. With the use of mobile technology instead of laptops, librarians could much more freely move through the areas of the lab where researchers are working in order to provide more efficient consultation services. In a survey completed after the implementation of this program, researchers commented on the convenience and time savings this service provided. Researchers who previously did not venture to the physical library space benefitted from the increased accessibility to librarians and library services.

In another outreach program at Sauls Memorial Library, which serves the Piedmont Hospital network at multiple locations, librarians physically travelled to outlying sites to support learning and clinical care by promoting library services and resources.[14] Although all locations have access to the library's website, employees and care providers expressed interest in having library services on-site. Librarians came up with a travelling program that brought a basic collection of materials on clinical, management, and nursing topics as well as general reading materials to each outreach site one day a month. The most popular material for each site proved to be audiobooks on CD. These materials were originally intended for patients but also became popular with staff members. Instead of transporting materials to and from the library, iPods could be loaded with multiple audio books and circulated on-site. Another popular collection included various consumer health–related books and videos, which could also be loaded onto mobile devices for circulation. Access to MedlinePlus and other resources could be mediated by librarians with tablet devices, who could then train remote users on using the library's website resources to meet their information needs.

FACILITATING LIBRARY COMMUNICATION WITH MOBILE DEVICES

Mobile technology has a huge potential to help facilitate library communication with internal community members as well as external stakeholders through outreach programs, as discussed above. Some specific applications include FaceTime, the web-conferencing tool built into iPads, and other web-conferencing tools with apps such as GoToMeeting, Join.me, Skype, and Google+ Hangouts. Social networking apps such as those for Facebook, Twitter, and LinkedIn can also facilitate communication in many settings. Mobile devices can be used to take photos and videos, which can be instantly emailed or shared via social networking apps. Document sharing between group members is also easy from mobile devices with apps that connect to cloud-based programs such as Dropbox, Google Drive, and SkyDrive. All of these tools can be used daily in libraries to communicate with library patrons, students, faculty, or other librarians at the point of need by using mobile devices.

At Boise State University, administrators felt that knowing how to use mobile devices was an important skill for all library employees.[15] In 2012, they purchased iPads for all library employees, regardless of job title. The "iPads for All" project allowed all employees to explore the use of mobile technology and become familiar with its capabilities. Staff members were encouraged to use the devices in whatever ways worked best for them professionally as well as personally. The project created a new respect and awareness among all staff members as they shared innovative and creative ways to use their iPads in their work. Beyond familiarizing all library staff with the use of mobile technology, the spirit of inclusiveness engendered by this project was found to be beneficial to the library work culture. Top reported reasons for iPads use were convenience, portability, and job-related tasks. Some of the ways employees commonly used iPads on the job included note-taking, researching questions that arise in meetings, locating items when in the stacks, eliminating the need for printing, answering reference questions when away from the desk, and using the device's camera to document physical evidence for maintenance.

Health sciences libraries are particularly impacted by the growth of mobile technology. The use of mobile technology in health, referred to as mHealth, is a fast-growing field.[16] In its simplest form, mHealth is done using SMS technology. This form of mHealth is being deployed especially in developing countries where infrastructure for wired networks is lacking, but most people have access to cell phone technology and infrastructure. SMS technology or texting is used to promote health among at-risk populations and to encourage treatment adherence. Text messages are sent to patients to educate and encourage them in their personal health management along with reminders to take medicine or perform other health-related tasks. For those with access to smartphones, the potential for mHealth is even greater. In 2012 it was estimated that there were over 40,000 health-related apps available, and the numbers are growing exponentially.[17] Health-related apps range from specialized health care apps for providers to apps for managing specific diseases or for personal health management. There are also devices that can connect with mobile devices to record health information, such as the popular Fitbit activity tracker. Concerns have been raised about the quality of many of the apps targeted at consumers. Quality regulation is lacking for mHealth apps. Librarians can help by vetting quality apps and recommending the appropriate ones to their community members. Librarians can also partner with public health workers and health care providers to help create quality mHealth applications and programs. More about this topic is discussed in chapter 10.

PROFESSIONAL DEVELOPMENT AND LIFELONG LEARNING

Librarians, as well as other academic and medical professionals, must complete professional development activities to keep up with new trends, retain certifications, work toward tenure requirements, and satisfy personal goals of lifelong learning. Mobile technology has the potential to significantly enhance professional development by providing access to online professional development opportunities at the most convenient places and times. A few minutes on the bus or while waiting for appointments can be used for productive professional development. Many professional development opportunities are now available in online modules that keep track of the user's progress so that he or she can work on modules in small increments until finished. Other professional development activities are available through online webinars, videos, and tutorials as well as through vendor-provided modules within library resources for medical and other personnel. Mobile technology can facilitate lifelong learning by expanding opportunities for formal learning into daily life and making it easy to engage with informal as well as formal learning activities as they arise.

A 2010 article in the *British Journal of Education* refers to learning with mobile technology as "seamless learning," and calls for international collaboration in developing mobile learning practices.[18] Much research still needs to be done to find the best ways to provide learning opportunities using mobile technology. With ubiquitous mobile technology, learning can become a constant activity, beyond the need for classroom spaces and set meeting times. Communities of learners can be established that reach beyond buildings, borders, and even continents. Using the personalization features of mobile technology, learners can have the flexibility to create an optimum environment for learning that is fully customized to their learning needs. Before we reach that utopian future, much work needs to be done to create standards for creation and assessment of online learning environments. Mobile technology has the potential to open up access to lifelong learning to everyone regardless of location, social position, or financial status. With further standardization of devices, cross-platform development practices, worldwide access to mobile infrastructure, and open policies on who can access the Internet, the dream of pervasive lifelong learning for all can become a reality.

NOTES

1. Wikipedia contributors. "Outreach." *Wikipedia,* n.d. http://en.wikipedia.org/w/index.php?title=Outreach&oldid=616850525 (accessed July 16, 2014).
2. Fama, Jane, Donna Berryman, Nancy Harger, Paul Julian, Nancy Peterson, Margaret Spinner, and Jennifer Varney. "Inside Outreach: A Challenge for Health Sciences Librarians." *Journal of the Medical Library Association* 93, no. 3 (2005): 327.
3. Kraemer, Elizabeth W., Dana J. Keyse, and Shawn V. Lombardo. "Beyond These Walls: Building a Library Outreach Program at Oakland University." *Reference Librarian* 39, no. 82 (2004): 5–17.
4. Covone, Nicole, and Mia Lamm. "Just Be There: Campus, Department, Classroom . . . and Kitchen?" *Public Services Quarterly* 6, nos. 2–3 (2010): 198–207.

5. Lee, Angela, and Ann Whitney Gleason. "Tablet Mania: Exploring the Use of Tablet Computers in an Academic Health Sciences Library." *Journal of Hospital Librarianship* 12, no. 3 (2012): 281–87.

6. Fabian, Carole Ann, Charles D'aniello, Cynthia Tysick, and Michael Morin. "Multiple Models for Library Outreach Initiatives." *Reference Librarian* 39, no. 82 (2004): 39–55.

7. Lee and Gleason, "Tablet Mania."

8. Kendrick, James. "iPad mini: Bringing the Elderly into the Digital Age." ZDNet, December 14, 2012. http://www.zdnet.com/ipad-mini-bringing-the-elderly-into-the-digital-age-7000008817/ (accessed July 23, 2014).

9. "iPads Help Older People Get More Out Of Life." MyAgeingParent, n.d. http://www.myageingparent.com/how-ipads-can-help-your-ageing-parent-get-more-out-of-life/ (accessed July 23, 2014).

10. Kouame, Gail, and David Young. "Promoting Health Literacy and Personal Health Management with Inmates in a County Detention Center." *Journal of Hospital Librarianship* 14, no. 2 (2014): 172–79.

11. Drewes, Kathy, and Nadine Hoffman. "Academic Embedded Librarianship: An Introduction." *Public Services Quarterly* 6, nos. 2–3 (2010): 75–82.

12. Covone and Lamm, "Just Be There."

13. Brandenburg, Marci D., Alan Doss, and Tracie E. Frederick. "Evaluation of a Library Outreach Program to Research Labs." *Medical Reference Services Quarterly* 29, no. 3 (2010): 249–59.

14. Lacy, Edie, and Sharon Leslie. "Library Outreach Near and Far: Programs to Staff and Patients of the Piedmont Healthcare System." *Medical Reference Services Quarterly* 26, no. 3 (2007): 91–103.

15. Armstrong, Michelle, and Peggy S. Cooper. "Experiencing "iPads for All": Results from a Library-wide Mobile Technology Program." Proceedings of the Charleston Library Conference, 2014. http://dx.doi.org/10.5703/1288284315320 (accessed July 23, 2014).

16. Boulos, Maged N. Kamel, Ann C. Brewer, Chante Karimkhani, David B. Buller, and Robert P. Dellavalle. "Mobile Medical and Health Apps: State of the Art, Concerns, Regulatory Control and Certification." *Journal of Public Health Informatics* 5, no. 3 (2014): 229.

17. Pelletier, Stephen G. "Explosive Growth in Health Care Apps Raises Oversight Questions." AAMC Reporter, October 2012. https://www-aamc-org.offcampus.lib.washington.edu/newsroom/reporter/october2012/308516/health-care-apps.html (accessed August 12, 2014).

18. Looi, Chee-Kit, Peter Seow, BaoHui Zhang, Hyo-Jeong So, Wenli Chen, and Lung-Hsiang Wong. "Leveraging Mobile Technology for Sustainable Seamless Learning: A Research Agenda." *British Journal of Educational Technology* 41, no. 2 (2010): 154–69.

RESOURCES

AnticoagEvaluator app (iOS and Android): https://itunes.apple.com/us/app/anticoagevaluator/id609795286?mt=8 and https://play.google.com/store/apps/details?id=org.acc.AnticoagEvaluator&hl=en

ARUP Consult app (iOS): http://www.arupconsult.com/

Browzine app (iOS and Android): http://thirdiron.com/browzine/

Calculate by QxMD app (iOS, Android and BlackBerry): http://www.qxmd.com/apps/calculate-by-qxmd

Dynamed app (iOS and Android, institutional subscription needed): https://dynamed.ebscohost.com/access/mobile

Micromedex app (iOS and Android, subscription needed): http://micromedex.com/mobile

Needham, Gill, and Mohamed Ally. *M-libraries 2: A Virtual Library in Everyone's Pocket*. London: Facet Publishing, 2010.

Peters, Thomas A., and Lori Bell, eds. *The Handheld Library: Mobile Technology and the Librarian*. Santa Barbara: CA: ABC-CLIO, 2013.

Radiology 2.0 app (iOS): https://itunes.apple.com/us/app/radiology-2.0-one-night-in/id397926581?mt=8

Read by QxMD app (iOS and Android): http://www.qxmd.com/apps/read-by-qxmd-app
Spectrum > Mobile Learning, Libraries, and Technologies: http://mobile-libraries.blogspot.com
VisualDX app (iOS and Android): http://www.visualdx.com/features/mobile-access

Chapter Ten

The Future of Mobile Technology

Imagine a future where the watch on your wrist tells you when to leave for your next appointment, the eyeglasses you wear record everything you see, a Band-Aid helps keep your blood sugar levels steady, and your home thermostat system automatically raises the temperature as you head home in your car. This future is actually a reality now. The Pebble watch is a tiny mobile device that can send emails, texts, news alerts, and much more with over 1,000 apps already available. Google Glass is a mobile device connected to your eyeglasses that can track your fitness, record activities, search and navigate, stream news feeds, and play music. The Nest thermostat can interact with a Mercedes vehicle to adjust your home temperature before you get there. Medical scientists are working on trials of a Band-Aid-sized patch that delivers insulin as needed, making it unnecessary for diabetics to check blood sugar and inject insulin. Mobile technology is rapidly spreading into every aspect of our lives. Virtual reality, wearable technology, mHealth, and the Internet of Things are some of the futuristic trends that are emerging from the mobilization of technology.

EMERGING TRENDS IN MOBILE TECHNOLOGY

The New Media Consortium's Horizon Report charts technology trends and challenges each year for K–12 and higher education sectors. Each year, the NMC maps out the top emerging trends and predicts future developments in technology that will affect the educational community in transformative ways. Mobile technology has been featured prominently on the Horizon Report for many years now. The recently released 2014 Horizon Report predicts further transformative effects from mobile technology, including many topics that have been discussed in the previous chapters of this book.[1] Trends that have already emerged, according to the Horizon Report, include online learning and increased use of social media in education. Both of these trends are facilitated by mobile technology. Another trend that is just emerging is a shift, due to our more mobile society, from passive consumers to creators of self-published creative works. This trend emerges in education as a move towards fostering more interactive learning environments where students participate in creating knowledge and take more ownership over their educational goals. Another trend that is starting to emerge in education is analyzing all the data stored online from web-based learning for better assessment of learning and personalized education for all students. A longer-term idea that is slowly emerging comes from the increasing use of online learning and improvements to learning management systems and accompanying technologies that will allow online learning to become more interactive through video and social networking technologies.

The Horizon Report also tracks developments in educational technology. Two trends that are being implemented currently are the flipped classroom and the use of learning analytics. The flipped classroom is the idea of students learning independently online, at their own pace, through watching videos of recorded lectures and reading e-content. Actual class time is used for group activities and project-based learning. The idea of the flipped classroom is enabled and greatly enhanced by the availability of mobile technology. Another emerging development, which is facilitated by mobile technology, is the collection of data from mobile, online course management systems that can be used to personalize and target educational experiences to individual students. A development that is starting to be more accepted in the educational area is the use of games for educational purposes. Serious games or "gamification" of traditional educational topics is becoming increasingly accepted as a valid means of engaging students with traditional subjects in a way that is creative and effective. Further out on the horizon of new technology is the idea of "quantified self" and the virtual assistant. As people integrate the use of mobile devices more and more in their daily lives through new wearable devices and tracking of personal metrics such as weight and exercise levels, large amounts of data are collected for each person. As mobile technology continues to advance with voice recognition, better location awareness, and gesture technology, so does the idea of virtual "assistants," which we interact with through our mobile devices to facilitate our daily tasks and even start to anticipate our every need.

The Horizon Report also discusses significant challenges that stand in the way of implementing new technologies. Some of the problems standing in the way of the adoption of these new technologies include the low digital literacy skills of many faculty members, and librarians are no exception. Training is needed to bring faculty skills to the level needed where they can successfully participate in a flipped classroom and facilitate learning with social media and online course management systems. Another challenge is actually the proliferation of online learning models such as massive online open courses (MOOC). Further study is needed in order to use these new models in ways that facilitate higher level learning in students. A huge challenge in educational systems today is expanding access to education and the pressure it puts on traditional educational models. Mobile technology has the potential to significantly increase access to education, especially to remote populations in underdeveloped countries who are increasingly gaining access to mobile technology. Further study is needed to create sustainable models and develop standards for providing increasingly mobile and independent learning that provides quality learning experiences as well as allowing for educating a larger quantity of students at the same time as is done with MOOCs.

The 2014 Horizon Report, Library Edition, reports on technology trends specifically driving change in libraries.[2] Chief among these trends is the prioritization of mobile educational content and delivery of that content. Because of research such as that conducted by the Pew Research Group, which shows that greater numbers of people increasingly own and use mobile technology, libraries are finding the need to adapt content and services to accommodate mobile users. Mobile-optimized access to library services, the use of mobile apps, e-books and e-readers, mobile devices for checkout in the library, and increasing mobile access to e-resources are all high priorities. There is a great need for policies and standardization to be developed for resources and search platforms. Problems of e-resource sharing and accessibility also need to be addressed and solved in order to move mobile priorities further along.

Mobile apps, smartphones, and tablet devices are trends already firmly established around the world, and libraries are no exception. Apps are inexpensive and typically do one specific thing, so mobile device users typically download and use many apps on their devices. Gartner predicts that by 2017, there will have been over 268 billion downloads of mobile apps, making this technology the top way to reach consumers with products and content (figure 10.1).[3] One of the reasons that apps are free or inexpensive is that the use of the app provides data to the app creator. Service providers such as Google, Apple, Facebook, and Amazon are gathering this data in order to predict future service and product needs and to provide personalized, directed advertising to consumers. Apps are being used in libraries today for a variety of services, such as reading e-resources and accessing databases. Many libraries publish lists of subject related apps and many now have their own personalized, custom library app, providing access to library services, catalogs, and library information to community members. Some libraries are beginning to explore creating apps in even more innovative ways such as custom search tools, games for library navigation and learning research skills, as well as apps using augmented reality and GPS to provide virtual tours of collections or historical objects.

The Horizon Report, Libraries Edition, also points out that one of the major challenges facing adoption of mobile technology by libraries is the need for radical change in order to keep up with the rapid technology changes happening in academic institutions as well as in the public arena. Leadership is needed to bring libraries into the next generation and beyond. Shifts in attitude are crucial in order to effect real change so that libraries are not just keeping up with change, but leading it for their institutions. Libraries need transformational leadership to identify innovative practices that can deal with constant technological change and lead their institutions in using that technology to enhance academic and public services to support institutional mission and goals. Further study to establish policies and guidelines to help libraries navigate these changes is critical.

The yearly Horizon Reports predict many examples of these emerging technologies, which are now becoming a reality in our everyday lives. Libraries and librarians are slowly embracing some of these technologies but remain suspicious of others. As the new technology becomes more mainstream in our society, libraries will need to be open to providing services that their patrons want and are becoming used to having in their daily lives. Some of the emerging mobile technology trends that libraries should watch for as they continue to be developed include gaming and virtual reality, wearable technology, mHealth, and the so-called Internet of Things.

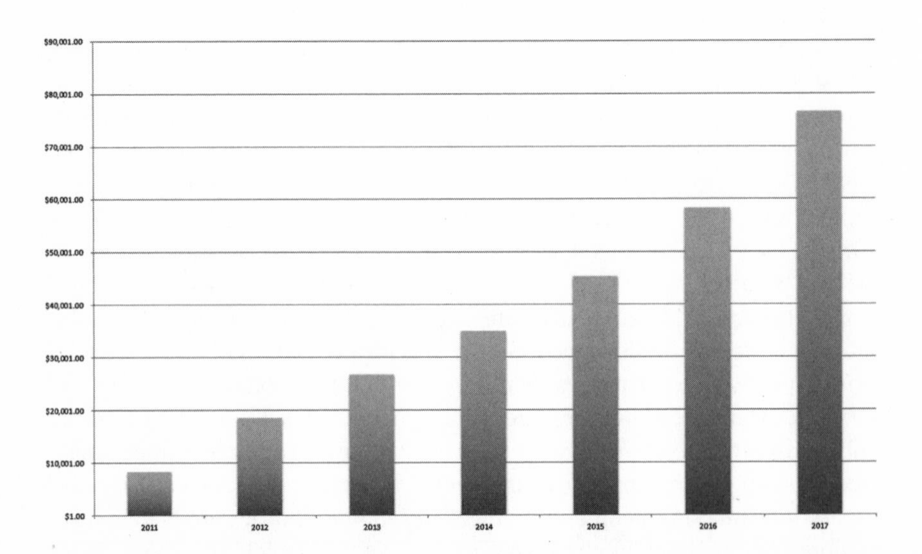

Figure 10.1. App sales by 2017. ***Gartner***

GAMING AND VIRTUAL REALITY

Games for mobile devices are extremely popular and many are available for all types of devices. Serious games or "gamification" enables educators to present content in an engaging and fun way so that students of all ages enjoy the process of learning. Entertainment apps include many that have historical or scientific content built in, such as the Natural History Museum's Evolution game for iPad and the Gravity Launch app. There are also many gaming apps available for use in the classroom, such as Simple Physics, the NASA app, and Minecraft: Pocket Edition. Librarians at Grand Valley State University created a library game called Library Quest, which works on both Android and iOS platforms.[4] They contracted with an outside app designer to build the game, in which students earn points when they use library services and spaces. Designing games is not an easy process, and it can be very expensive to hire a designer. Keeping the game simple and focused on predetermined outcomes is important to building a successful product. Designing the game on paper and performing usability testing before investing in a consultant to do the game coding is crucial. It is important to remember that games are supposed to be fun, so the focus should be on the player experience, not just the content.

Virtual reality (VR) or augmented reality (AR) is a new and exciting way to provide information interactively. Augmented reality happens when images, video, or audio information is virtually attached to a real-world location or object such as a building, artwork, or photograph.[5] The technology behind augmented reality has been around in the sciences and the military for many years, but is moving into the mainstream due to increasing use with mobile technology. There are a growing number of augmented reality apps appearing for both Android and iOS devices. Much AR development has been focused on gaming, but new software is emerging that allows anyone to create location-based content and share it with others through apps that use the camera and GPS functions of mobile devices. Layar, Wikitude, and Aurasma are a few of the leading AR software companies. With AR software, a user can scan a location or object and attach content such as a video, text, or images and then publish to the web through one of the AR software applications. Then, when someone using one of the many available AR apps scans the object with a mobile device, the added content is displayed. Wikitude World Browser is a popular free app that displays AR content from thousands of providers such as Yelp and Trip Advisor.[6] Some popular AR apps for iOS and Android include Acrossair AR browser, Tagwhat, iOnRoad, SpotCrime, and Google Sky Map. An AR app for Windows phone users is called Nokia City Lens.

At Stetson University, librarians have experimented with using AR in the law library, using the free Stiktu application that was available from Layar.[7]

Stiktu allowed a user to scan a physical object and attach informational content to be uploaded and saved on Stiktu servers. Anyone with the Stiktu app could scan the object with their mobile device and view the AR content, as well as add to the content if desired. Stiktu could also publish to social networking apps such as Facebook and Twitter in order to publicize the AR content. Unfortunately, this easy-to-use app has been discontinued, but other AR apps can produce the same results. Icons can be placed at locations where AR content is available in order to prompt people to use the app, just like QR codes are used today. Some ideas for the use of AR in the library include scavenger hunts, labeling of library locations, and virtual reference. AR technology is just getting started, and limitations include difficulties with scanning objects exactly right so that the virtual content displays correctly, issues with small cameras and low-resolution screens on mobile devices, and problems with getting content to display correctly on multiple types of devices. A drawback to this technology in the future could be the posting of derogatory or offensive "graffiti" to virtual spaces, so some filtering of postings may be needed if this technology becomes more commonplace than it is today.

WEARABLE TECHNOLOGY

The next generation of mobile technology is emerging as wearable technology. This innovation combines tiny sensors with existing mobile technologies such as apps, wireless, Bluetooth and wi-fi, and GPS. The result is an interactive way to go about everyday activities such as exercising, sleeping, eating, and communicating with friends and coworkers. Some uses for wearable technology in libraries might be hands-free, wireless connecting to online information while moving about the library, remotely controlling presentations while moving around the room working with students, and receiving calendar notifications and email instantly and everywhere you go during your day. Health sciences librarians are particularly challenged to keep up with the rising trend in wearable devices for monitoring health and fitness. As wearable technology becomes more mainstream, more and better apps will be developed, making these devices even more innovative. We will discuss some of the more popular devices, including Google Glass, the Pebble watch, the Myo armband, and the Fitbit activity tracker.

Google Glass is a controversial wearable technology. Currently in beta, Google Glass is obtainable only by being recommended by a current user of Glass and is very expensive, severely limiting the number and types of users who have access to this technology. Google Glass is basically a mobile phone that you wear like glasses. It contains a camera and microphone to record what you see and hear as well as displaying sound and video information from the web, including augmented reality content, directly to your eyes and ears. The display is mounted on glasses frames slightly above your line of sight so that you can choose to ignore the display or look up to see information, messages, or travel directions. Google Glass was first released in 2013 to selected "Glass Explorers."[8] This limited marketing campaign allows Google to selectively test the beta device. Since so few can actually obtain one, the device is not well understood and issues around privacy arise when people suspect that they are being photographed or videotaped without their permission by those wearing Glass. Some librarians have embraced this new technology and have been invited by others to purchase the device, resulting in Glass lending programs where academic community members are able to check out the device from the library to give it a try and participate in an ongoing community review of the device.[9]

The Pebble "smart" watch is the first to look and feel like wearing a normal wristwatch without being heavy and awkward. The Pebble allows you to choose from many decorative watch faces, and the wristband also comes in several bright colors. The Pebble works with both iOS and Android phones via a Bluetooth connection. The battery reportedly lasts about a week before needing to be recharged, and the watch is fully waterproof. Out of the box, it displays the time like a regular watch, but the key is to install apps that connect to your smartphone in order to display incoming phone calls and messages. With Pebble, you can check for incoming calls and check email messages by just looking down at the watch face rather than having to check your phone. It also sends calendar notifications and Facebook messages. The power of the Pebble watch is that new apps are being developed to allow it to connect to social networking sites, track fitness and health, play games, and even remotely control your phone. The Pebble apps install on your phone and work through the Bluetooth connection between the watch and your phone. These devices could eventually replace the tablet computers or smartphones of today, making people even more mobile and making it an even greater priority for librarians and libraries to become more virtual- and mobile-friendly. Apple releases the Apple Watch in early 2015, and it remains to be seen how popular this new device becomes.

New gesture-based technology is due to be released in September 2014. Myo is a black armband that fits around your arm just below the elbow, and extends gesture technology from the pinch and tap used with smartphones and tablets to the use of your entire arm. The Myo armband has sensors that measure electrical activity from your muscles in order to detect specific gestures made with your hand and forearm. The website video (http://www.thalmic.com/en/myo) shows a teacher controlling a PowerPoint presentation with just the movement of his arm and gestures of his hand. This device has obvious uses in the educational world to control presentation equipment from a distance in order to focus on interactions with students. Combined with display technology, this technology could develop into the scenario portrayed in the movie *Minority Report*, where massive amounts of information are displayed on multiple screens and controlled by hand gestures in order to sift through information to find connections and patterns in the data displayed.

The Fitbit wristband is one of several brands of wearable technology designed to track physical activity. This technology goes beyond just counting steps taken in daily activity toward weight management, food logging, and sleep optimization. The Fitbit is basically a pedometer that counts steps, but can also sync wirelessly with a mobile device as well as a computer to save activity levels, and keep food and sleep logs for a complete health outlook. Fitbit also works with social networking sites such as Facebook to post your activity levels so you can compare and compete with friends and encourage and motivate each other. Fitbit has many apps available, including fitness programs and monitors, weight management programs that connect to digital scales and sleep calculators, as well as utility and entertainment apps. Fitbit apps convert your mobile device into a health and wellness monitor. The Aria "smart" scale was created with built-in wi-fi so that weight, BMI, and body fat statistics can be uploaded to your mobile device or computer to track health data. Fitbit works with iOS, Android, and Windows Phone devices. There is now an app to connect Fitbit to Microsoft HealthVault, a free service that allows you to store health data online.

MHEALTH

Dr. Eric Topol, in his TED Talk, "The Wireless Future of Medicine," predicts that wireless technology will allow people to check all of their most important vital signs via smartphones and specialized apps (figure 10.2).[10] This technology works by means of a sensor located on the body that wirelessly sends signals to a smartphone or the Internet via cloud applications. Much of this technology is already available. Some of the new medical wireless technology includes Band-Aid-sized sensors implanted under the skin to continuously monitor blood sugar levels, and sensors that measure and save heart rate and other vital signs to monitor heart failure patients. Mobile ultrasound using wireless technology is also available, which could be used by individuals to monitor pregnancy or check for body changes that could indicate cancer. Continuous monitoring of hypertension, sleep, diet and activity, as well as monitoring for drug levels for such diseases as depression and monitoring the location of Alzheimer's patients have the potential to ensure maximum health for all.

Besides revolutionizing health care, mHealth devices have the potential to revolutionize medical research as well. In a recent article in *Scientific American*, the author states that the National Institutes of Health is currently

Ten top targets for mHealth		
Disease	# Affected	mHealth apps
Obesity	80 M	Smart scales, glucose, caloric intake, activity
Hypertension	74 M	Continuous BP, med compliance
Sleep Disorders	40 M	Sleep phases, quality, apnea, vitals
Diabetes	24 M	Glucose, hemoglobin, A1C
Asthma	23 M	RR, FEV1, Air quality, oximetry, pollen counts
Depression	21 M	Med Compliance, activity, communication
COPD	10 M	RR, FEV1, air quality, oximetry
Alzheimer's	5 M	Vitals, location, activity, balance
Heart Failure	5 M	Cardiac pressures, weight, BP, fluid status
Breast Cancer	3 M	Ultrasound self-exam save to web

Figure 10.2. Top targets for wireless medicine. *Topol TEDTalk*

working on establishing a research base for assessing the risks or benefits of these devices.[11] mHealth devices and sensors are capable of storing continuous health statistical data from patients over long periods of time. This technology has the potential to significantly change medical research. Patients can be fitted with these devices and continuously monitored remotely without much impact to their daily lives. Clinical trial research could be greatly enhanced with the use of remote monitoring devices, making health measurements more precise and reliable. Massive amounts of data will be collected because of the increased use of these devices, which may lead to breakthroughs in disease control and in finding underlying causes of disease. Librarians could be extremely valuable partners with mHealth researchers who need help with the classification and evaluation of mHealth devices, along with the management and organization of the large volumes of data they produce.

THE INTERNET OF THINGS

Prior to the American Library Association's annual conference in June 2014, a symposium on the Internet of Things (IOT) was hosted by the Online Computer Library Center (OCLC).[12] Featuring Daniel Obodovski, coauthor of a recent book on IOT, the symposium initiated a heated discussion on the dangers of collecting personal data that is shared on the Internet. The Internet of Things is made possible by the introduction of Internet Protocol version 6 (IPv6), the advanced system of assigning IP addresses that increases the number of addresses that can be assigned to over 600 trillion.[13] That is enough unique addresses to assign to all the devices we use in everyday life in order to connect them to the Internet and make them "smart" devices.

These devices may be connected to your body, as was described above in the section on mHealth, or they may be connected to each other, such as devices that monitor your home and automatically adjust temperature when they connect with your car as you approach your home. Obodovski feels that libraries have much to gain from gathering patron data in order to customize services to individual users. With the Internet of Things, libraries could automatically recommend books to patrons via their smartphones by accessing their reading history and search data. Other applications could include automatically tracking where people go and the services they use most often in a library, automatically suggesting research papers that may be applicable to individual research projects, and providing data on successful uses of library resources as opposed to failed or abandoned searches. Whether or not the usefulness of the IOT outweighs the commercialization and exploitation of personal data collected on the Internet remains to be seen. Librarians could be instrumental in working with data researchers to create policies and safeguards for personal data collection.

WHAT WILL 2020 BRING?

As the world becomes increasingly mobilized, what will the world of the future look like? Will libraries become obsolete, replaced by Google and Wikipedia, or will they change and adapt to the new information landscape and thrive? As search technology and data collection continues to grow at exponential rates, and everyone in the world becomes connected to the Internet, finding quality information and answering research questions will become increasingly difficult. The role of libraries throughout time has been the collection of information, and libraries will continue to collect information from all over the world, but it is also important to make sense of the information that is collected and make it retrievable by everyone, no matter where they are located or what language they speak. While information brokering will become more and more an online experience, physical libraries will still be places for community members to gather. Libraries that focus on the user experience and meeting the unique needs of the communities they serve will not only survive, but will continue to thrive in 2020 and beyond.

NOTES

1. New Media Consortium. "Horizon Higher Education Preview." NMC, 2014. http://www.nmc.org/pdf/2014-horizon-he-preview.pdf (accessed August 25, 2014).
2. New Media Consortium. "Horizon Report Library Edition." NMC, 2014. http://cdn.nmc.org/media/2014-nmc-horizon-report-library-EN.pdf (accessed August 25, 2014).
3. "Gartner Says by 2017, Mobile Users Will Provide Personalized Data Streams to More than 100 Apps and Services Every Day." Gartner, January 22, 2014. http://www.gartner.com/newsroom/id/2654115 (accessed August 31, 2014).

4. Felker, Kyle. "Library Quest: Developing a Mobile Game App for a Library." ACRL TechConnect Blog, September 17, 2013. http://acrl.ala.org/techconnect/?p=3783 (accessed August 31, 2014).

5. Cassella, Dena. "What Is Augmented Reality (AR): Augmented Reality Defined, iPhone Augmented Reality Apps and Games and More." Digital Trends, November 3, 2009. http://www.digitaltrends.com/mobile/what-is-augmented-reality-iphone-apps-games-flash-yelp-android-ar-software-and-more/#!bNBDd6 (accessed August 31, 2014).

6. Widder, Brandon. "Best Augmented Reality Apps." Digital Trends, March 14, 2014. http://www.digitaltrends.com/mobile/best-augmented-reality-apps/#!bNBHF5 (accessed August 31, 2014).

7. Barnes, Elizabeth, and Robert M. Brammer. "Bringing Augmented Reality to the Academic Law Library." *AALL Spectrum* 17 (2012): 13.

8. Labash, Matt. "Through a Google Glass, Darkly." *Weekly Standard*, April 28, 2014. http://www.weeklystandard.com/articles/through-google-glass-darkly_787020.html (accessed August 31, 2014).

9. Signorelli, Paul. "OK, Glass." *American Libraries Magazine*, January 26, 2014. http://www.americanlibrariesmagazine.org/blog/ok-glass (accessed August 31, 2014).

10. " Eric Topol: The Wireless Future of Medicine." YouTube video, 16:58, posted by TED Talks, February 23, 2010. https://www.youtube.com/watch?v=pTZM9X3JfTk (accessed August 31, 2014).

11. Collins, Francis. "The Real Promise of Mobile Health Apps." *Scientific American*. Vol. 307, July 1, 2012. http://www.scientificamerican.com/article/real-promise-mobile-health-apps/ (accessed August 31, 2014).

12. Pera, Mariam. "Libraries and the 'Internet of Things.'" *American Libraries Magazine*, June 28, 2014. http://www.americanlibrariesmagazine.org/blog/libraries-and-internet-things (accessed August 31, 2014).

13. Hahn, Jim. "The Internet of Things Meets the Library of Things." ACRL TechConnect Blog, March 19, 2012. http://acrl.ala.org/techconnect/?p=474 (accessed August 31, 2014).

RESOURCES

Wearable Technology

Fitbit: http://www.fitbit.com/
Google Glass: http://www.google.com/glass/start/
Myo: https://www.thalmic.com/en/myo/
Pebble: https://getpebble.com/

Gaming

Gravity Launch app (iPad and Android): http://sciencenetlinks.com/tools/gravity-launch-app/
Minecraft: Pocket Edition (iOS and Android): https://minecraft.net/pocket
NASA app (iOS and Android): http://www.nasa.gov/centers/ames/iphone/index.html
NHM: Evolution (iPad): http://www.nhm.ac.uk/business-centre/publishing/books/evolution/evolution-app/evolution-app.html
SimplePhysics (iOS and Android): https://itunes.apple.com/us/app/simplephysics/id408233979?mt=8 and https://play.google.com/store/apps/details?id=com.andrewgarrison.simplephysics&hl=en

Augmented Reality

Acrossair AR browser (iOS): https://itunes.apple.com/us/app/acrossair-augmented-reality/id348209004?mt=8

Aurasma: http://www.aurasma.com/
Google Sky Map (Android): https://play.google.com/store/apps/details?id=com.google.android.stardroid&hl=en
iOnRoad (iOS and Android): http://www.ionroad.com/
Layar: https://www.layar.com/
Nokia City Lens (Windows Phone): http://www.windowsphone.com/en-us/store/app/nokia-city-lens/93301a45-5849-4aad-a68e-c7c95df83ca1
SpotCrime (iOS and Android): https://itunes.apple.com/us/app/spotcrime+/id767693374?mt=8 and https://play.google.com/store/apps/developer?id=SpotCrime&hl=en
Tagwhat (iOS and Android): http://www.tagwhat.com/
Wikitude: http://www.wikitude.com/

Internet of Things

Nest: https://nest.com/

Index

About the Author

Ann Whitney Gleason currently serves as the head of the Health Sciences Library at Stony Brook University on Long Island, New York. Previously, she was associate director for Resources and Systems at the University of Washington Health Sciences Library (HSL), and served as the head of Computer Systems for the HSL. While at the University of Washington, she also served as liaison to the School of Social Work. Earlier in her career, Gleason worked as an educational technology specialist, technology director, and CIO for several years before receiving her MLIS from the University of Rhode Island. She also holds a BA in education. Gleason's research interests are currently focused on the use of educational technology in libraries, specifically mobile technologies, course management systems, and educational gaming.